#1199

TENDENCIES

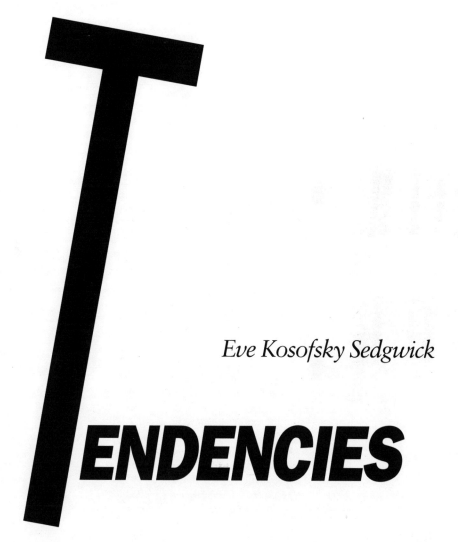

Eve Kosofsky Sedgwick

TENDENCIES

London 1994

ROUTLEDGE

First published 1994
by Routledge
11 New Fetter Lane, London EC4P 4EE

© 1993 Duke University Press
Typeset in Sabon by Tseng Information Systems

British Library Cataloguing in Publication Data
A catalogue record for this book is available from the
British Library
ISBN 0-415-10814-4   0-415-10815-2 (pbk)

In memory of Michael Lynch,

and with love to him.

CONTENTS

FOREWORD

T TIMES

At the 1992 gay pride parade in New York City, there was a handsome, intensely muscular man in full leather regalia, sporting on his distended chest a T-shirt that read, KEEP YOUR LAWS OFF OF MY UTERUS.

The two popular READ MY LIPS T-shirts marketed by ACT UP were also in evidence, and by the thousands. But for the first time it was largely gay men who were wearing the version of the shirt that features two turn-of-the-century-looking women in a passionate clinch. Most of the people wearing the version with the osculating male sailors, on the other hand, were lesbians. FAGGOT and BIG FAG were the T-shirt legends self-applied by many, many women; DYKE and the more topical LICK BUSH by many, many men.

(In a 1992 pre-Christmas ad for a popular gay T-shirt maker, the DYKE shirt is modeled by a man. The FAG HAG shirt is modeled by a man in drag.)

And everywhere at the march, on women and on men, there were T-shirts that said simply: QUEER.

It was a QUEER time.

It *feels* queer, and good—I'm sure I wasn't the only one at the march to have this sensation—when the wave of a broadly based public movement somehow overtakes and seems to amplify (amplifies by drowning it out?) the little, stubborn current of an individual narrative or obsession, an individual wellspring of narrow, desiring cathexis and cognition. On the scene of national gay/lesbian activism, in the *Village Voice,* in the 'zines, on the streets and even in some classrooms, I suppose this must be called the moment of Queer. (*Though it's other moments, too. Right now: the moment just before the inauguration of an ostentatiously heterosexual president who makes an audible, persistent claim to support lesbian and gay rights. Long moment of a deathly silence that means the AIDS drugs we've been struggling to hold on for are just not in the pipeline. When Melvin Dixon and Tom Yingling disappear from us, and Audre Lorde. When, for the first time, the New York pride march has massive participation by African Americans and Latinos. When people organize around claiming the label bisexual; or the right to be lesbian and gay soldiers; or the right to have their children* not *protected against HIV transmission.*)

In the short-shelf-life American marketplace of images, maybe the queer moment, if it's here today, will for that very reason be gone tomorrow. But I mean the essays collected in this book to make, cumulatively, stubbornly, a counterclaim against that obsolescence: a claim that something about *queer* is inextinguishable. Queer is a continuing moment, movement, motive—recurrent, eddying, *troublant*. The word "queer" itself means *across*—it comes from the Indo-European root *-twerkw*, which also yields the German *quer* (transverse), Latin *torquere* (to twist), English *athwart*. Titles and subtitles that at various times I've attached to the essays in *Tendencies* tend toward "across" formulations: *across genders, across sexualities, across genres, across "perversions."* (The book itself would have had an "across" subtitle, but I just couldn't choose.) The *queer* of these essays is transitive—multiply transitive. The immemorial current that *queer* represents is antiseparatist as it is antiassimilationist. Keenly, it is relational, and strange.

Most of these essays were concurrent with my work on another book, *Epistemology of the Closet* (1990). *Epistemology* made two hypotheses about the transitivity of sexual identities, but it was really organized around only one of them; the other is the concern of most of the *Tendencies* essays. I hypothesized in *Epistemology of the Closet* that modern homo/heterosexual definition has become so exacerbated a cultural site because of an enduring incoherence about whether it is to be thought of as an issue only for a minority of (distinctly lesbian or gay) individuals, or instead as an issue that cuts across every locus of agency and subjectivity in the culture. The other hypothesis proffered in *Epistemology*, though, but not much explored there, is that modern sexual identities are also structured by a double-binding but immensely productive incoherence about gender. Specifically, I suggest in *Epistemology* that current "common sense" about homosexuality involves two contradictory gender models: forms of the inversion trope (in which a gay man represents "a woman's soul trapped in a man's body," and so on) that see queer people as uniquely situated *between* genders; and forms of the trope of gender separatism, under which it seems only common sense

> that people of the same gender, people grouped together under the single most determinate diacritical mark of social organization, people whose economic, institutional, emotional, physical needs and knowledges may have so much in common, should bond together also on the axis of sexual desire. . . . Gender-separatist models would thus place the woman-loving woman and the man-loving man each at the "natural" defining center of their own gender, again in contrast to inversion models that locate gay people—whether biologically or culturally—at the threshold between genders.[1]

Like *Epistemology of the Closet,* the essays in *Tendencies* do not aim to adjudicate these conflictual dispensations. But by various means a lot of the essays attempt to find new ways to think about lesbian, gay, and other sexually dissident loves and identities in a complex social ecology where the presence of different genders, different identities and identifications, will be taken as a given. I have no use at all for the trope of homosexuality-as-gender inversion, with its heterosexist presumption that *only* a self that is somehow "really a man" could be attracted to a woman, or vice versa. Yet the T-shirted carnival of cross-reference at last summer's pride parade also evokes a history of moments to which the gender separatist models just won't answer.

These essays are about passionate queer things that happen across the lines that divide genders, discourses, and "perversions." What does it mean that Denis Diderot, in *The Nun,* makes lesbian knowledge the synecdoche for all knowledge? That Willa Cather refracts lesbian desires through gay male identities? That for Henry James the crucial term "ironic" seems to have been an epithet that referred to lesbian sexuality? That John Waters's movies insistently stage the condensation of fat women with gay men? Or that *Sense and Sensibility* originates a modern novelistic subjectivity in the space between a masturbating girl and the women and men who love her?

These are essays in a variety of modes, but several of them also have an experimental ambition in common. I've wanted to recruit—but also where I could, to denude or somehow transfigure—the energies of some received

1. Eve Kosofsky Sedgwick, *Epistemology of the Closet* (Berkeley: University of California Press, 1990), pp. 87–88.

forms of writing that were important to me: the autobiographical narra-tive, the performance piece, the atrocity story, the polemic, the prose essay that quotes poetry, the obituary. There's a lot of first person singular in this book (and some people hate that), and it's there for different reasons in different essays; to begin with, though, I'd find it mutilating and disin-genuous to disallow a grammatical form that marks the site of such dense, accessible effects of knowledge, history, revulsion, authority, and pleasure. Perhaps it would be useful to say that the first person throughout repre-sents neither the sense of a simple, settled congratulatory "I," on the one hand, nor on the other a fragmented postmodernist postindividual—never mind an unreliable narrator. No, "I" is a heuristic; maybe a powerful one.

Tendencies ought to make sense to people who haven't read *Episte-mology of the Closet;* but the intimate adhesion between the books meant that I couldn't afford to be too embarrassed about some few pages of overlap. Changes from previously published versions of these essays have mostly been to eliminate places where two essays cover the same ground. I've indulged myself by smoothing out some prose that was already setting my teeth on edge, too; but I haven't tried to override the individual essay structure of these pieces, nor to make them sound as if they were all writ-ten out of a single moment or impulse or analysis.

Two of these essays were written while I was a Rockefeller Fellow at the National Humanities Center. I wish the rest of the world were as solici-tously arranged to help people accomplish the work that's important to them! Like most universities, Duke isn't quite; but it has been terrific in supporting this inquiry, and *Tendencies* owes a great deal to both Duke's generosity and that of Amherst College.

I suppose there's something a bit louche about it, the way the acknowledg-ment convention lets one savor the pride of being indebted to wonderful friends. *Next book,* I tell myself, I'll shut up about the *mishpocheh.* But why? These paragraphs are about pleasure, not propriety.

The writing of *Tendencies* was almost coterminous with that of *Episte-mology of the Closet*—with a couple of extra years added on to the end—so the acknowledgments to that book can also stand, with big second help-ings of love to the same patient creditors, as the acknowledgments to this. It's said to be a commonplace among people who have been seriously ill that you never come out of an episode of illness with the same friends you went into it with. My experience over the past two years has been different from that: there have been treasured new friends, but then most sustainingly there have been all the treasured old ones. I've especially looked forward to saying here how much I depended, during the first stages of a cancer diagnosis and a pretty draining half-year of chemotherapy, on the energy, cheer, and love brought to North Carolina in visits from Henry Abelove, Mary Campbell, Robert Dawidoff, Joe Gordon, Janet Halley, Andy Parker, Cindy Patton, Libby Potter, Mary Russo, Nancy Waring, and Carolyn Williams. Tim Gould offered the very finest in long-distance counseling. Judy Frank, Sasha Torres, Richard Fung, Margie Waller, David Robinson, Michèle Aina Barale, Judith Butler helped in some inimitable ways.

It didn't require this firsthand experience of illness to sharpen my sense of privilege in the friendship of a variety of men and women constrained by debility or love to deal closely with HIV—in their own bodies, as care-givers, as activists. But my illness has strengthened and, for me, qualita-tively changed many of these relationships. Simon Watney, to let one such friend stand for many, is a challenge as he is an inspiration, his accomplish-ment representable only in the silent calculus of lives unlost.

A few more people to mention. For years now I've been peculiarly lucky in the matter of fairy godmothers; they've included Jim Kincaid, Helen Vendler, Barbara Herrnstein Smith, Neil Hertz, and Liz Clark, and I thank them for many acts of magic. Songmin Shin has marvelously become one of the Kosofskys whose stimulus and affection are a constant tonic to this work. Ken Wissoker is as good an editor as he is a friend, which is going some; I've had excellent support from previous editors, but working with Ken on Series Q has quite changed my sense of what that relationship can accomplish. (Yes, as a matter of fact this *is* a plug to prospective series authors.) Shannon Van Wey kindly maintains my head at the proper small size. Rita Kosofsky gives me a motive for these projects, as her teaching has given me means for them. And the ineffably comforting Julian Olive is the only person who generally knows if I'm coming or going.

Either Jonathan Flatley or Gustavus Stadler came up with the title for this book. (I know *Tendencies* sounds like a title for Walter Benjamin essays—really it's channeled from Dame Edna Everage). They're among the present and recent Duke graduate students—José Muñoz, Brian Sel-sky, Adam Frank, Mary Armstrong, Graham Hammill are only a few of the others—who have created something unusual here: a supported and chal-lenging space for original queer thought. It's been wonderful to have access to it. Katie Kent and Mandy Berry are two whose invaluable contributions to this book (editorial, critical, emotional) have been even more direct.

Finally, there are my dear companions of this time, Jonathan Gold-berg, Hal Sedgwick, and Michael Moon. The tenderest care—most incisive thought—most uncompromising reading: it's hard to say which aspects of Jon's friendship are farthest beyond price. No praise is more validating than his, no severity more instructive, no kindness more penetrating. I've always known about Hal the same quite unlikely thing that, in Proust, the narrator knows about his grandmother: "that everything that was mine, my cares, my wishes, would be . . . supported upon a desire to save and prolong my life stronger than was my own." Living with Michael is a virtually nonstop Scheherazade party. Underlying his brilliance is an outrageous gift of *being interested* in the whole world and soliciting the cathexis of others for it. His coauthorship of one of the essays in this book, important though it has been to me, can reflect only a small part of the wealth I've experienced in his intimacy.

The simple secret of the plot is just to tell them that I love them a lot.

QUEER

AND

NOW

A MOTIVE   I think everyone who does gay
and lesbian studies is haunted by the suicides
of adolescents. To us, the hard statistics come
easily: that queer teenagers are two to three
times likelier to attempt suicide, and to accomplish it, than others; that up to 30 percent of teen
suicides are likely to be gay or lesbian; that a third
of lesbian and gay teenagers say they have attempted
suicide; that minority queer adolescents are at even
more extreme risk.[1]

The knowledge is indelible, but not astonishing, to
anyone with a reason to be attuned to the profligate way
this culture has of denying and despoiling queer energies
and lives. I look at my adult friends and colleagues doing
lesbian and gay work, and I feel that the survival of each one
is a miracle. Everyone who survived has stories about how it
was done

—an outgrown anguish
Remembered, as the Mile

Our panting Ankle barely passed—
When Night devoured the Road—
But we—stood whispering in the House—
And all we said—was "Saved"!

1. Paul Gibson, "Gay Male and Lesbian Youth Suicide," U.S. Department of Health and
Human Services, *Report of the Secretary's Task Force on Youth Suicide* (Washington, D.C.,
1989), vol. 3, pp. 110–142.

(as Dickinson has it).[2] How to tell kids who are supposed never to learn this, that, farther along, the road widens and the air brightens; that in the big world there are worlds where it's plausible, our demand to *get used to it*.

EPISTEMOLOGIES I've heard of many people who claim they'd as soon their children were dead as gay. What it took me a long time to believe is that these people are saying no more than the truth. They even speak for others too delicate to use the cruel words. For there is all the evidence. The preponderance of school systems, public and parochial, where teachers are fired, routinely, for so much as intimating the right to existence of queer people, desires, activities, children. The routine denial to sexually active adolescents, straight *and* gay, of the things they need—intelligible information, support and respect, condoms—to protect themselves from HIV transmission. (As a policy aimed at punishing young gay people with death, this one is working: in San Francisco for instance, as many as 34 percent of the gay men under twenty-five being tested—and 54 percent of the young black gay men—are now HIV infected.)[3] The systematic separation of children from queer adults; their systematic sequestration from the truth about the lives, culture, and sustaining relations of adults they know who may be queer. The complicity of parents, of teachers, of clergy, even of the mental health professions in invalidating and hounding kids who show gender-dissonant tastes, behavior, body language. In one survey 26 percent of young gay men had been forced to leave home because of conflicts with parents over their sexual identity;[4] another report concludes that young gays and lesbians, many of them throwaways, comprise as many as a quarter of all homeless youth in the United States.[5]

And adults' systematic denial of these truths to ourselves. The statistics on the triple incidence of suicide among lesbian and gay adolescents come from a report prepared for the U.S. Department of Health and Human Services in 1989; under congressional pressure, recommendations based on this section of the report were never released. Under congressional pres-

---

2. *The Complete Poems of Emily Dickinson*, ed. Thomas H. Johnson (Boston: Little, Brown, 1960), poem 325, p. 154.

3. T. A. Kellogg et al., "Prevalence of HIV-1 Among Homosexual and Bisexual Men in the San Francisco Bay Area: Evidence of Infection Among Young Gay Men," *Seventh International AIDS Conference Abstract Book*, vol. 2 (Geneva, 1991) (W.C. 3010), p. 298.

4. G. Remafedi, "Male Homosexuality: The Adolescent's Perspective," unpublished manuscript, Adolescent Health Program, University of Minnesota, 1985. Cited in Gibson, "Gay Male and Lesbian Youth Suicide."

5. Gibson, "Gay Male and Lesbian Youth Suicide," pp. 113–15.

sure, in 1991 a survey of adolescent sexual behavior is defunded. Under the threat of congressional pressure, support for all research on sexuality suddenly (in the fall of 1991) dries up. Seemingly, this society wants its children to know nothing; wants its queer children to conform or (and this is not a figure of speech) die; and wants not to know that it is getting what it wants.

PROMISING, SMUGGLING, READING, OVERREADING This history makes its mark on what, individually, we are and do. One set of effects turns up in the irreducible multilayeredness and multiphasedness of what queer survival means—since being a survivor on this scene is a matter of surviving *into* threat, stigma, the spiraling violence of gay- and lesbian-bashing, and (in the AIDS emergency) the omnipresence of somatic fear and wrenching loss. It is also to have survived into a moment of unprecedented cultural richness, cohesion, and assertiveness for many lesbian and gay adults. Survivors' guilt, survivors' glee, even survivors' responsibility: powerfully as these are experienced, they are also more than complicated by how permeable the identity "survivor" must be to the undiminishing currents of risk, illness, mourning, and defiance.

Thus I'm uncomfortable generalizing about people who do queer writing and teaching, even within literature; but some effects do seem widespread. I think many adults (and I am among them) are trying, in our work, to keep faith with vividly remembered promises made to ourselves in childhood: promises to make invisible possibilities and desires visible; to make the tacit things explicit; to smuggle queer representation in where it must be smuggled and, with the relative freedom of adulthood, to challenge queer-eradicating impulses frontally where they are to be so challenged.

I think that for many of us in childhood the ability to attach intently to a few cultural objects, objects of high or popular culture or both, objects whose meaning seemed mysterious, excessive, or oblique in relation to the codes most readily available to us, became a prime resource for survival. We needed for there to be sites where the meanings didn't line up tidily with each other, and we learned to invest those sites with fascination and love. This can't help coloring the adult relation to cultural texts and objects; in fact, it's almost hard for me to imagine another way of coming to care enough about literature to give a lifetime to it. The demands on both the text and the reader from so intent an attachment can be multiple, even paradoxical. For me, a kind of formalism, a visceral near-identification with the writing I cared for, at the level of sentence structure, metrical pattern, rhyme, was one way of trying to appropriate what seemed the numinous and resistant power of the chosen objects. Education made it easy to accumulate tools for this particular formalist project, because the texts that magnetized me happened to be novels and poems; it's impressed me deeply the way others of my generation and since seem to have invented for themselves, in the spontaneity of great need, the tools for a formalist apprehension of other less prestigious, more ubiquitous kinds of text: genre movies, advertising, comic strips.

For me, this strong formalist investment didn't imply (as formalism is generally taken to imply) an evacuation of interest from the passional, the imagistic, the ethical dimensions of the texts, but quite the contrary: the need I brought to books and poems was hardly to be circumscribed, and I felt I knew I would have to struggle to wrest from them sustaining news of the world, ideas, myself, and (in various senses) my kind. The reading practice founded on such basic demands and intuitions had necessarily to run against the grain of the most patent available formulae for young people's reading and life—against the grain, often, of the most accessible voices even in the texts themselves. At any rate, becoming a perverse reader was never a matter of my condescension to texts, rather of the surplus charge of my trust in them to remain powerful, refractory, and exemplary. And this doesn't seem an unusual way for ardent reading to function in relation to queer experience.

WHITE NIGHTS The first lesbian and gay studies class I taught was in the English Department at Amherst College in 1986. I thought I knew which five or six students (mostly queer) would show up, and I designed the course, with them in mind, as a seminar that would meet one evening a week, at my house. The first evening sixty-five students showed up—a majority of them, straight-identified.

Having taught a number of these courses by now, I know enough to expect to lose plenty of sleep over each of them. The level of accumulated urgency, the immediacy of the demand that students bring to them, is jolting. In most of their courses students have, unfortunately, learned to relinquish the expectation that the course material will address them where they live and with material they can hold palpably accountable; in gay/lesbian courses, though, such expectations seem to rebound, clamorous and unchastened, in all their rawness. Especially considering the history of denegation that most queer students bring with them to college, the vitality of their demand is a precious resource. Most often during a semester everyone will spend some time angry at everybody else. It doesn't surprise me when straight and gay students, or women and men students, or religious and nonreligious students have bones to pick with each other or with me.

What has surprised me more is how divisive issues of methodology and disciplinarity are: the single most controversial thing in several undergraduate classes has been *that they were literature courses,* that the path to every issue we discussed simply had to take the arduous defile through textual interpretation.

Furthermore, it was instructive to me in that class at Amherst that a great many students, students who defined themselves as nongay, were incensed when (in an interview in the student newspaper) I told the story of the course's genesis. What outraged them was the mere notation that I had designed the course evisioning an enrollment of mostly lesbian and gay students. Their sense of entitlement as straight-defined students was so strong that they considered it an inalienable right to have all kinds of different lives, histories, cultures unfolded as if anthropologically in formats specifically designed—designed from the ground up—for maximum legibility to themselves: they felt they shouldn't so much as have to slow down the Mercedes to read the historical markers on the battlefield. That it was a field where the actual survival of other people in the class might at the very moment be at stake—where, indeed, in a variety of ways so might their own be—was hard to make notable to them among the permitted assumptions of their liberal arts education. Yet the same education was being used so differently by students who brought to it sharper needs, more supple epistemological frameworks.

CHRISTMAS EFFECTS   What's "queer"? Here's one train of thought about it. The depressing thing about the Christmas season—isn't it?—is that it's the time when all the institutions are speaking with one voice. The Church says what the Church says. But the State says the same thing: maybe not (in some ways it hardly matters) in the language of theology, but in the language the State talks: legal holidays, long school hiatus, special postage stamps, and all. And the language of commerce more than chimes in, as consumer purchasing is organized ever more narrowly around the final weeks of the calendar year, the Dow Jones aquiver over Americans' "holiday mood." The media, in turn, fall in triumphally behind the Christmas phalanx: ad-swollen magazines have oozing turkeys on the cover, while for the news industry every question turns into the Christmas question—Will hostages be free *for Christmas?* What did that flash flood or mass murder (umpty-ump people killed and maimed) do to those families' *Christmas?* And meanwhile, the pairing "families/Christmas" becomes increasingly tautological, as families more and more constitute themselves according to the schedule, and in the endlessly iterated image, of the holiday itself constituted in the image of "the" family.

The thing hasn't, finally, so much to do with propaganda for Christianity as with propaganda for Christmas itself. They all—religion, state, capital, ideology, domesticity, the discourses of power and legitimacy—line up with each other so neatly once a year, and the monolith so created is a thing one can come to view with unhappy eyes. What if instead there were a practice of valuing the ways in which meanings and institutions can be at loose ends with each other? What if the richest junctures weren't the ones where *everything means the same thing?* Think of that entity "the family," an impacted social space in which all of the following are meant to line up perfectly with each other:

a surname
a sexual dyad
a legal unit based on state-regulated marriage
a circuit of blood relationships
a system of companionship and succor
a building
a proscenium between "private" and "public"
an economic unit of earning and taxation
the prime site of economic consumption
the prime site of cultural consumption
a mechanism to produce, care for, and acculturate children
a mechanism for accumulating material goods over several generations
a daily routine
a unit in a community of worship
a site of patriotic formation

and of course the list could go on. Looking at my own life, I see that—probably like most people—I have valued and pursued these various elements of family identity to quite differing degrees (e.g., no use at all for worship, much need of companionship). But what's been consistent in this particular life is an interest in *not* letting very many of these dimensions line up directly with each other at one time. I see it's been a ruling intuition for me that the most productive strategy (intellectually, emotionally) might be, whenever possible, to *dis*articulate them one from another, to *dis*engage them—the bonds of blood, of law, of habitation, of privacy, of companionship and succor—from the lockstep of their unanimity in the system called "family."

Or think of all the elements that are condensed in the notion of sexual identity, something that the common sense of our time presents as a unitary category. Yet, exerting any pressure at all on "sexual identity," you see that its elements include

your biological (e.g., chromosomal) sex, male or female;

your self-perceived gender assignment, male or female (supposed to be the same as your biological sex);

the preponderance of your traits of personality and appearance, masculine or feminine (supposed to correspond to your sex and gender);

the biological sex of your preferred partner;

the gender assignment of your preferred partner (supposed to be the same as her/his biological sex);

the masculinity or femininity of your preferred partner (supposed to be the opposite[6] of your own);

your self-perception as gay or straight (supposed to correspond to whether your preferred partner is your sex or the opposite);

your preferred partner's self-perception as gay or straight (supposed to be the same as yours);

your procreative choice (supposed to be yes if straight, no if gay);

your preferred sexual act(s) (supposed to be insertive if you are male or masculine, receptive if you are female or feminine);

your most eroticized sexual organs (supposed to correspond to the procreative capabilities of your sex, and to your insertive/receptive assignment);

your sexual fantasies (supposed to be highly congruent with your sexual practice, but stronger in intensity);

your main locus of emotional bonds (supposed to reside in your preferred sexual partner);

your enjoyment of power in sexual relations (supposed to be low if you are female or feminine, high if male or masculine);

the people from whom you learn about your own gender and sex (supposed to correspond to yourself in both respects);

your community of cultural and political identification (supposed to correspond to your own identity);

---

6. The binary calculus I'm describing here depends on the notion that the male and female sexes are each other's "opposites," but I do want to register a specific demurral against that bit of easy common sense. Under no matter what cultural construction, women and men are more like each other than chalk is like cheese, than ratiocination is like raisins, than up is like down, or than 1 is like 0. The biological, psychological, and cognitive attributes of men overlap with those of women by vastly more than they differ from them.

and—again—many more. Even this list is remarkable for the silent presumptions it has to make about a given person's sexuality, presumptions that are true only to varying degrees, and for many people not true at all: that everyone "has a sexuality," for instance, and that it is implicated with each person's sense of overall identity in similar ways; that each person's most characteristic erotic expression will be oriented toward another person and not autoerotic; that if it is alloerotic, it will be oriented toward a single partner or kind of partner at a time; that its orientation will not change over time.[7] Normatively, as the parenthetical prescriptions in the list above suggest, it should be possible to deduce anybody's entire set of specs from the initial datum of biological sex alone—if one adds only the normative assumption that "the biological sex of your preferred partner" will be the opposite of one's own. With or without that heterosexist assumption, though, what's striking is the number and *difference* of the dimensions that "sexual identity" is supposed to organize into a seamless and univocal whole.

And if it doesn't?

That's one of the things that "queer" can refer to: the open mesh of possibilities, gaps, overlaps, dissonances and resonances, lapses and excesses of meaning when the constituent elements of anyone's gender, of anyone's sexuality aren't made (or *can't be* made) to signify monolithically. The experimental linguistic, epistemological, representational, political adventures attaching to the very many of us who may at times be moved to describe ourselves as (among many other possibilities) pushy femmes, radical faeries, fantasists, drags, clones, leatherfolk, ladies in tuxedoes, feminist women or feminist men, masturbators, bulldaggers, divas, Snap! queens, butch bottoms, storytellers, transsexuals, aunties, wannabes, lesbian-identified men or lesbians who sleep with men, or . . . people able to relish, learn from, or identify with such.

Again, "queer" can mean something different: a lot of the way I have used it so far in this dossier is to denote, almost simply, same-sex sexual object choice, lesbian or gay, whether or not it is organized around multiple criss-crossings of definitional lines. And given the historical and contemporary force of the prohibitions against *every* same-sex sexual expression, for anyone to disavow those meanings, or to displace them from the term's definitional center, would be to dematerialize any possibility of queerness itself.

At the same time, a lot of the most exciting recent work around "queer"

7. A related list that amplifies some of the issues raised in this one appears in the introduction to *Epistemology of the Closet*, pp. 25–26.

spins the term outward along dimensions that can't be subsumed under gender and sexuality at all: the ways that race, ethnicity, postcolonial nationality criss-cross with these *and other* identity-constituting, identity-fracturing discourses, for example. Intellectuals and artists of color whose sexual self-definition includes "queer"—I think of an Isaac Julien, a Gloria Anzaldúa, a Richard Fung—are using the leverage of "queer" to do a new kind of justice to the fractal intricacies of language, skin, migration, state. Thereby, the gravity (I mean the *gravitas*, the meaning, but also the *center* of gravity) of the term "queer" itself deepens and shifts.

Another telling representational effect. A word so fraught as "queer" is—fraught with so many social and personal histories of exclusion, violence, defiance, excitement—never can only denote; nor even can it only connote; a part of its experimental force as a speech act is the way in which it dramatizes locutionary position itself. Anyone's use of "queer" about themselves means differently from their use of it about someone else. This is true (as it might also be true of "lesbian" or "gay") because of the violently different connotative evaluations that seem to cluster around the category. But "gay" and "lesbian" still present themselves (however delusively) as objective, empirical categories governed by empirical rules of evidence (however contested). "Queer" seems to hinge much more radically and explicitly on a person's undertaking particular, performative acts of experimental self-perception and filiation. A hypothesis worth making explicit: that there are important senses in which "queer" can signify only *when attached to the first person*. One possible corollary: that what it takes—all it takes—to make the description "queer" a true one is the impulsion *to* use it in the first person.

CURRENT: PROJECT 1 *The Golden Bowl,* J. L. Austin, *Dr. Susan Love's Breast Book,* and Mme de Sévigné are stacked up, open-faced, on the chair opposite me as I write. I've got three projects braiding and unbraiding in my mind; the essays in this book mark, in different ways, their convergent and divergent progress. I see them as an impetus into future work as well.

Project 1—most of the essays in this book embody it—is about desires and identifications that move across gender lines, including the desires of men for women and of women for men. In that sense, self-evidently, heterosexuality is one of the project's subjects. But the essays are queer ones. Their angle of approach is directed, not at reconfirming the self-evidence and "naturalness" of heterosexual identity and desire, but rather at rendering those culturally central, apparently monolithic constructions newly accessible to analysis and interrogation.

The project is difficult partly because of the asymmetries between the speech relations surrounding heterosexuality and homosexuality. As Michel Foucault argues, during the eighteenth and nineteenth centuries in Europe,

> Of course, the array of practices and pleasures continued to be referred to [heterosexual monogamy] as their internal standard; but it was spoken of less and less, or in any case with a growing moderation. Efforts to find out its secrets were abandoned; nothing further was demanded of it than to define itself from day to day. The legitimate couple, with its regular sexuality, had a right to more discretion. It tended to function as a norm, one that was stricter, perhaps, but quieter. . . .
>
> Although not without delay and equivocation, the natural laws of matrimony and the immanent rules of sexuality began to be recorded on two separate registers.[8]

Thus, if we are receptive to Foucault's understanding of modern sexuality as the most intensive site of the demand for, and detection or discursive production of, the Truth of individual identity, it seems as though this silent, normative, uninterrogated "regular" heterosexuality may not function as a sexuality at all. Think of how a culturally central concept like public/private is organized so as to preserve for heterosexuality the unproblematicalness, the apparent naturalness, of its *discretionary* choice between display and concealment: "public" names the space where cross-sex couples *may,* whenever they feel like it, display affection freely, while same-sex couples *must* always conceal it; while "privacy," to the degree that it is a right codified in U.S. law, has historically been centered on the protection-from-scrutiny of the married, cross-sex couple, a scrutiny to which (since the 1986 decision in *Bowers* v. *Hardwick*) same-sex relations on the other hand are unbendingly subject. Thus, heterosexuality is consolidated as the *opposite* of the "sex" whose secret, Foucault says, "the obligation to conceal . . . was but another aspect of the duty to admit to."[9] To the degree that heterosexuality does not function as a sexuality, however, there are stubborn barriers to making it accountable, to making it so much as visible, in the framework of projects of historicizing and hence denaturalizing sexuality. The making historically visible of heterosexuality is difficult because, under its institutional pseudonyms such as Inheritance, Marriage, Dynasty,

8. Michel Foucault, *The History of Sexuality*, vol. 1: *An Introduction*, trans. Robert Hurley (New York: Pantheon Books, 1978), pp. 38–40.
9. Ibid., p. 61.

Family, Domesticity, and Population, heterosexuality has been permitted to masquerade so fully as History itself—when it has not presented itself as the totality of Romance.

PROJECT 2 Here I'm at a much earlier stage, busy with the negotiations involved in defining a new topic in a usable, heuristically productive way; it is still a series of hunches and overlaps; its working name is Queer Performativity. You can see the preoccupations that fuel the project already at work throughout *Epistemology of the Closet* as well as *Tendencies,* but I expect it to be the work of a next book to arrive at broadly usable formulations about them. Like a lot of theorists right now (Judith Butler and her important book *Gender Trouble* can, perhaps, stand in for a lot of the rest of us), I'm interested in the implications for gender and sexuality of a tradition of philosophical thought concerning certain utterances that do not merely describe, but actually perform the actions they name: "*J'accuse*"; "Be it resolved . . ."; "I thee wed"; "I apologize"; "I dare you." Discussions of linguistic performativity have become a place to reflect on ways in which language really can be said to produce effects: effects of identity, enforcement, seduction, challenge.[10] They also deal with how powerfully language *positions:* does it change the way we understand meaning, for instance, if the semantic force of a word like "queer" is so different in a first-person from what it is in a second- or third-person sentence?

My sense is that, in a span of thought that arches at least from Plato to Foucault, there are some distinctive linkages to be traced between linguistic performativity and histories of same-sex desire. I want to go further with an argument implicit in *Epistemology of the Closet:* that both the act of coming out, and closetedness itself, can be taken as dramatizing certain features of linguistic performativity in ways that have broadly applicable implications. Among the striking aspects of considering closetedness in this framework, for instance, is that the speech act in question is a series of silences! I'm the more eager to think about performativity, too, because it may offer some ways of describing what *critical* writing can effect (promising? smuggling?); anything that offers to make this genre more acute and experimental, less numb to itself, is a welcome prospect.

10. One of the most provocative discussions of performativity in relation to literary criticism is Shoshana Felman, *The Literary Speech Act: Don Juan with J. L. Austin, or Seduction in Two Languages,* trans. Catherine Porter (Ithaca, N.Y.: Cornell University Press, 1983); most of the current work being done on performativity in relation to sexuality and gender is much indebted to Judith Butler's *Gender Trouble: Feminism and the Subversion of Identity* (New York: Routledge, 1989).

PROJECT 3   This project involves thinking and writing about something that's actually structured a lot of my daily life over the past year. Early in 1991 I was diagnosed, quite unexpectedly, with a breast cancer that had already spread to my lymph system, and the experiences of diagnosis, surgery, chemotherapy, and so forth, while draining and scary, have also proven just sheerly *interesting* with respect to exactly the issues of gender, sexuality, and identity formation that were already on my docket. (Forget the literal-mindedness of mastectomy, chemically induced menopause, etc.: I would warmly encourage anyone interested in the social construction of gender to find some way of spending half a year or so as a totally bald woman.) As a general principle, I don't like the idea of "applying" theoretical models to particular situations or texts—it's always more interesting when the pressure of application goes in both directions—but all the same it's hard not to think of this continuing experience as, among other things, an adventure in applied deconstruction.[11] How could I have arrived at a more efficient demonstration of the instability of the supposed oppositions that structure an experience of the "self"?—the part and the whole (when cancer so dramatically corrodes that distinction); safety and danger (when fewer than half of the women diagnosed with breast cancer display any of the statistically defined "risk factors" for the disease); fear and hope (when I feel—I've got a quarterly physical coming up—so much less prepared to deal with the news that a lump or rash *isn't* a metastasis than that it is); past and future (when a person anticipating the possibility of death, and the people who care for her, occupy temporalities that more and more radically diverge); thought and act (the words in my head are aswirl with fatalism, but at the gym I'm striding treadmills and lifting weights); or the natural and the technological (what with the exoskeleton of the bone-scan machine, the uncanny appendage of the IV drip, the bionic implant of the Port-a-cath, all in the service of imaging and recovering my "natural" healthy body in the face of its spontaneous and endogenous threat against itself). Problematics of undecidability present themselves in a new, unfacile

11. That deconstruction can offer crucial resources of thought for survival under duress will sound astonishing, I know, to anyone who knows it mostly from the journalism on the subject —journalism that always depicts "deconstructionism," not as a group of usable intellectual tools, but as a set of beliefs involving a patently absurd dogma ("nothing really exists"), loopy as Christian Science but as exotically aggressive as (American journalism would also have us find) Islam. I came to my encounter with breast cancer not as a member of a credal sect of "deconstructionists" but as someone who needed all the cognitive skills she could get. I found, as often before, that I had some good and relevant ones from my deconstructive training.

way with a disease whose very *best* outcome—since breast cancer doesn't respect the five-year statute of limitations that constitutes cure for some other cancers—will be decades and decades of free-fall interpretive panic.

Part of what I want to see, though, is what's to be learned from turning this experience of dealing with cancer, in all its (and my) marked historical specificity, and with all the uncircumscribableness of the turbulence and threat involved, back toward a confrontation with the theoretical models that have helped me make sense of the world so far. The phenomenology of life-threatening illness; the performativity of a life threatened, relatively early on, by illness; the recent crystallization of a politics explicitly oriented around grave illness: exploring these connections *has* (at least for me it has) to mean hurling my energies outward to inhabit the very farthest of the loose ends where representation, identity, gender, sexuality, and the body can't be made to line up neatly together.

It's probably not surprising that gender is so strongly, so multiply valenced in the experience of breast cancer today. Received wisdom has it that being a breast cancer patient, even while it is supposed to pose unique challenges to one's sense of "femininity," nonetheless plunges one into an experience of almost archetypal Femaleness. Judith Frank is the friend whom I like to think of as Betty Ford to my Happy Rockefeller—the friend, that is, whose decision to be public about her own breast cancer diagnosis impelled me to the doctor with my worrisome lump; she and her lover, Sasha Torres, are only two of many women who have made this experience survivable for me: compañeras, friends, advisors, visitors, students, lovers, correspondents, relatives, caregivers (these being anything but discrete categories). Some of these are indeed people I have come to love in feminist- and/or lesbian-defined contexts; beyond that, a lot of the knowledge and skills that keep making these women's support so beautifully apropos derive from distinctive feminist, lesbian, and women's histories. (I'd single out, in this connection, the contributions of the women's health movement of the 70s—its trenchant analyses, its grass-roots and antiracist politics, its publications,[12] the attitudes and institutions it built and some of the careers it seems to have inspired.)

12. The work of this movement is most available today through books like the Boston Women's Health Book Collective's *The New Our Bodies, Ourselves: Updated and Expanded for the Nineties* (New York: Simon and Schuster, 1992). An immensely important account of dealing with breast cancer in the context of feminist, antiracist, and lesbian activism is Audre Lorde, *The Cancer Journals,* 2d ed. (San Francisco: Spinsters Ink, 1988) and *A Burst of Light* (Ithaca, N.Y.: Firebrand Books, 1988).

At the same time, though, another kind of identification was plaited inextricably across this one—not just for me, but for others of the women I have been close to as well. Probably my own most formative influence from a quite early age has been a viscerally intense, highly speculative (not to say inventive) cross-identification with gay men and gay male cultures as I inferred, imagined, and later came to know them. It wouldn't have required quite so overdetermined a trajectory, though, for almost any forty year old facing a protracted, life-threatening illness in 1991 to realize that the people with whom she had perhaps most in common, and from whom she might well have most to learn, are people living with AIDS, AIDS activists, and others whose lives had been profoundly reorganized by AIDS in the course of the 1980s.

As, indeed, had been my own life and those of most of the people closest to me. "Why me?" is the cri de coeur that is popularly supposed to represent Everywoman's deepest response to a breast cancer diagnosis—so much so that not only does a popular book on the subject have that title, but the national breast cancer information and support hotline is called Y-ME! Yet "Why me?" was not something it could have occurred to me to ask in a world where so many companions of my own age were already dealing with fear, debilitation, and death. I wonder, too, whether it characterizes the responses of the urban women of color forced by violence, by drugs, by state indifference or hostility, by AIDS and other illnesses, into familiarity with the rhythms of early death. At the time of my diagnosis the most immediate things that were going on in my life were, first, that I was coteaching (with Michael Moon) a graduate course in queer theory, including such AIDS-related material as Cindy Patton's stunning *Inventing AIDS*. Second, that we and many of the students in the class, students who indeed provided the preponderance of the group's leadership and energy at that time, were intensely wrapped up in the work (demonstrating, organizing, lobbying) of a very new local chapter of the AIDS activist organization ACT UP. And third, that at the distance of far too many miles I was struggling to communicate some comfort or vitality to a beloved friend, Michael Lynch, a pioneer in gay studies and AIDS activism, who seemed to be within days of death from an AIDS-related infection in Toronto.

"White Glasses," the final essay in *Tendencies*, tells more about what it was like to be intimate with this particular friend at this particular time. More generally though, the framework in which I largely experienced my diagnosis—and the framework in which my friends, students, house sharers, life companion, and others made available to me almost over-

whelming supplies of emotional, logistical, and cognitive sustenance[13]—
was very much shaped by AIDS and the critical politics surrounding it, including the politics of homophobia and of queer assertiveness. The AIDS
activist movement, in turn, owes much to the women's health movement
of the 70s; and in another turn, an activist politics of breast cancer, spear-
headed by lesbians, seems in the last year or two to have been emerging
based on the model of AIDS activism.[14] The dialectical epistemology of the
two diseases, too—the kinds of secret each has constituted; the kinds of
*out*ness each has required and inspired—has made an intimate motive for
me. As "White Glasses" says,

> It's as though there were transformative political work to be done just
> by being available to be identified with in the very grain of one's illness
> (which is to say, the grain of one's own intellectual, emotional, bodily
> self as refracted through illness and as resistant to it)—being available
> for identification to friends, but as well to people who don't love one;
> even to people who may not like one at all nor even wish one well.

MY WAR AGAINST WESTERN CIVILIZATION   That there were such people—
that, indeed, the public discourse of my country was increasingly dominated by them—got harder and harder to ignore during the months of my
diagnosis and initial treatment. For the first time, it was becoming routine
to find my actual name, and not just the labels of my kind, on those journalistic lists of who was to be considered more dangerous than Saddam Hus-
sein. In some ways, the timing of the diagnosis couldn't have been better:
if I'd needed a reminder I had one that, sure enough, life *is* too short, at
least mine is, for going head-to-head with people whose highest approbation, even, would offer no intellectual or moral support in which I could
find value. Physically, I was feeling out of it enough that the decision to let
this journalism wash over me was hardly a real choice—however I might
find myself misspelled, misquoted, mis-paraphrased, or (in one hallucinatory account) married to Stanley Fish. It was the easier to deal psychically

13. And physical: I can't resist mentioning the infallibly appetite-provoking meals that Jonathan Goldberg, on sabbatical in Durham, planned and cooked every night during many queasy
months of my chemotherapy.
14. On this, see Alisa Solomon, "The Politics of Breast Cancer," *Village Voice* 14 May 1991,
pp. 22–27; Judy Brady, ed., *1 in 3: Women with Cancer Confront an Epidemic* (Pittsburgh and
San Francisco: Cleis Press, 1991); Midge Stocker, ed., *Cancer as a Women's Issue: Scratching
the Surface* (Chicago: Third Side Press, 1991); and Sandra Butler and Barbara Rosenblum,
Cancer in Two Voices (San Francisco: Spinsters, 1991).

with having all these journalists scandalize my name because it was clear most of them wouldn't have been caught dead reading my work: the essay of mine that got the most free publicity, "Jane Austen and the Masturbating Girl," did so without having been read by a single one of the people who invoked it: it reached its peak of currency in hack circles months before it was published, and Roger Kimball's *Tenured Radicals,* which first singled it out for ridicule, seems to have gone to press before the essay was so much as *written.*[15]

Not that I imagine a few cozy hours reading *Epistemology of the Closet* would have won me rafts of fans amongst the punditterati. The attacks on me personally were based on such scummy evidential procedures that the most thin-skinned of scholars—so long as her livelihood was secure—could hardly have taken them to heart; the worst of their effects on me at the time was to give an improbable cosmic ratification (yes, actually, everything *is* about me!) to the self-absorption that forms, at best, an unavoidable feature of serious illness. If the journalistic hologram bearing my name seemed a relatively easy thing to disidentify from, though, I couldn't help registering with much greater intimacy a much more lethal damage. I don't know a gentler way to say it than that at a time when I've needed to make especially deep draughts on the reservoir of a desire to live and thrive, that resource has shown the cumulative effects of my culture's wasting depletion of it. It *is* different to experience from the vantage point of one's own bodily illness and need, all the brutality of a society's big and tiny decisions, explicit and encoded ones, about which lives have or have not value. Those decisions carry not only institutional and economic but psychic and, I don't doubt, somatic consequences. A thousand things make it impossible to mistake the verdict on queer lives and on women's lives, as on the lives of those who are poor or are not white. The hecatombs of queer youth; a decade squandered in a killing inaction on AIDS; the rapacious seizure from women of our defense against forced childbirth; tens of millions of adults and children excluded from the health care economy; treatment of homeless people as unsanitary refuse to be dealt with by periodic "sweeps"; refusal of condoms in prisons, persecution of needle exchange programs; denial and trivialization of histories of racism; or merely the pivot of a disavowing pronoun in a newspaper editorial: such things as these are facts, but at the same time they are piercing or murmuring voices in the heads

15. Roger Kimball, *Tenured Radicals: How Politics Has Corrupted Our Higher Education* (New York: Harper and Row, 1990), pp. 145–46.

of those of us struggling to marshal "our" resources against illness, dread, and devaluation. They speak to us. They have an amazing clarity.

A CRAZY LITTLE THING CALLED *RESSENTIMENT* There was something especially devastating about the wave of anti-"PC" journalism in the absolutely open contempt it displayed, and propagated, for *every* tool that has been so painstakingly assembled in the resistance against these devaluations. Through raucously orchestrated, electronically amplified campaigns of mock-incredulous scorn, intellectual and artistic as well as political possibilities, skills, ambitions, and knowledges have been laid waste with a relishing wantonness. No great difficulty in recognizing those aspects of the anti-"PC" craze that are functioning as covers for a rightist ideological putsch; but it has surprised me that so few people seem to view the recent developments as, among other things, part of an overarching history of anti-intellectualism: anti-intellectualism left as well as right. No twentieth-century political movement, after all, can afford not to play the card of populism, whether or not the popular welfare is what it has mainly at heart (indeed, perhaps especially where it is least so). And anti-intellectual pogroms, like anti-Semitic or queer-bashing ones, are quick, efficient, distracting, and almost universally understood signifiers for a populist solidarity that may boil down to nothing by the time it reaches the soup pot. It takes care and intellectual scrupulosity to forge an egalitarian politics not founded on such telegraphic slanders. Rightists today like to invoke the threatening specter of a propaganda-ridden socialist realism, but both they and the anti-intellectuals of the left might meditate on why the Nazis' campaign against "degenerate art" (Jewish, gay, modernist) was couched, as their own arguments are, in terms of assuring the instant, unmediated, and universal accessibility of all the sign systems of art (Goebbels even banning all art criticism in 1936, on the grounds that art is self-explanatory). It's hard to tell which assumption is more insultingly wrong: that the People (always considered, of course, as a monolithic unit) have no need and no faculty for engaging with work that is untransparent; or that the work most genuinely expressive of the People would be so univocal and so limpidly vacant as quite to obviate the labors and pleasures of interpretation. Anti-intellectuals today, at any rate, are happy to dispense with the interpretive process and depend instead on appeals to the supposedly self-evident: legislating against "*patently* offensive" art (no second looks allowed); citing titles as if they were texts; appealing to potted summaries and garbled trots as if they were variorum editions in the original Aramaic.

The most self-evident things, as always, are taken—as if unanswerably—
to be the shaming risibility of any form of oblique or obscure expression;
and the flat inadmissability of openly queer articulation.

THOUGHT AS PRIVILEGE   These histories of anti-intellectualism cut across
the "political correctness" debate in complicated ways. The term "politically correct" originated, after all, in the mockery by which experimen-
tally and theoretically minded feminists, queers, and leftists (of every color,
class, and sexuality) fought back against the stultifications of feminist and
left anti-intellectualism. The hectoring, would-be-populist derision that
difficult, ambitious, or sexually charged writing today encounters from the
right is not always very different from the reception it has already met with
from the left. It seems as if many academic feminists and leftists must be
grinding their teeth at the way the right has willy-nilly conjoined their discursive fate with that of theorists and "deconstructionists"—just as, to be
fair, many theorists who have betrayed no previous interest in the politics
of class, race, gender, or sexuality may be more than bemused at turning
up under the headings of "Marxism" or "multiculturalism." The right's
success in grouping so many, so contestative, movements under the rubric
"politically correct" is a coup of cynical slovenliness unmatched since the
artistic and academic purges of Germany and Russia in the thirties.

What the American intellectual right has added to this hackneyed populist semiotic of *ressentiment* is an iridescent oilslick of elitist self-regard.
Trying to revoke every available cognitive and institutional affordance
for reflection, speculation, experimentation, contradiction, embroidery,
daring, textual aggression, textual delight, double entendre, close reading,
free association, wit—the family of creative activities that might, for purposes of brevity, more simply be called *thought*—they yet stake their claim
as the only inheritors, defenders, and dispensers of a luscious heritage of
thought that most of them would allow to be read only in the dead light of
its pieties and its exclusiveness. Through a deafeningly populist rhetoric,
they advertise the mean pleasures of ranking and gatekeeping as available
to all. But the gates that we are invited to invigorate ourselves by cudgeling
barbarians at open onto nothing but a *Goodbye, Mr. Chips* theme park.

What is the scarcity that fuels all this *ressentiment?* The leveraged burn-
out of the eighties certainly took its toll, economically, on universities as
well as on other professions and industries. In secretaries' offices, in hos-
pitals and HMOs, in network news bureaus, in Silicon Valley laboratories
and beyond, the bottom line has moved much closer to a lot of people's
work lives—impinging not just on whether they *have* work, but on what

they do when they're there. But academic faculty, in our decentralized institutions, with our relatively diffuse status economy and our somewhat archaic tangle of traditions and prerogatives, have had, it seems, more inertial resistance to offer against the wholesale reorientation of our work practices around the abstractions of profit and the market. For some faculty at some colleges and universities, it is still strikingly true that our labor is divided up by task orientation (we may work on the book till it's done, explain to the student till she understands) rather than by a draconian time discipline; that what we produce is described and judged in qualitative as much as quantitative terms; that there is a valued place for affective expressiveness, and an intellectually productive permeability in the boundaries between public and private; that there are opportunities for collaborative work; and most importantly, that we can expend some substantial part of our paid labor on projects we ourselves have conceived, relating to questions whose urgency and interest make a claim on our own minds, imaginations, and consciences.

Millions of people today struggle to carve out—barely, at great cost to themselves—the time, permission, and resources, "after work" or instead of decently-paying work, for creativity and thought that will not be in the service of corporate profit, nor structured by its rhythms. Many, many more are scarred by the prohibitive difficulty of doing so. No two people, no two groups would make the same use of these resources, furthermore, so that no one can really pretend to be utilizing them "for" another. I see that some must find enraging the spectacle of people for whom such possibilities are, to a degree, built into the structure of our regular paid labor. Another way to understand that spectacle, though, would be as one remaining form of insistence that it is not inevitable—it is not a simple fact of nature—for the facilities of creativity and thought to represent rare or exorbitant *privilege*. Their economy should not and need not be one of scarcity.

The flamboyance with which some critical writers—I'm one of them— like to laminate our most ambitious work derives something, I think, from this situation. Many people doing all kinds of work are able to take pleasure in aspects of their work; but something different happens when the pleasure is not only taken but openly displayed. I like to make that different thing happen. Some readers identify strongly with the possibility of a pleasure so displayed; others disidentify from it with violent repudiations; still others find themselves occupying less stable positions in the circuit of contagion, fun, voyeurism, envy, participation, and stimulation. When the pleasure is attached to meditative or artistic productions that deal, not always in an effortlessly accessible way, with difficult and painful realities among others, then readers' responses become even more complex and dramatic, more productive for the author and for themselves. Little wonder then that sexuality, the locus of so many showy pleasures and untidy identities and of so much bedrock confrontation, opacity, and loss, should bear so much representational weight in arguments about the structure of intellectual work and life. Sexuality in this sense, perhaps, can *only* mean queer sexuality: so many of us have the need for spaces of thought and work where everything doesn't mean the same thing!

So many people of varying sexual practices, too, enjoy incorrigibly absorbing imaginative, artistic, intellectual, and affective lives that have been richly nourished by queer energies—and that are savagely diminished when the queerness of those energies is trashed or disavowed. In the very first of the big "political correctness" scare pieces in the mainstream press, *Newsweek* pontificated that under the reign of multiculturalism in colleges, "it would not be enough for a student to refrain from insulting homosexuals . . . . He or she would be expected to . . . study their literature and culture alongside that of Plato, Shakespeare, and Locke." [16] *Alongside?* Read any Sonnets lately? You dip into the *Phaedrus* often?

To invoke the utopian bedroom scene of Chuck Berry's immortal *aubade:* Roll over, Beethoven, and tell Tchaikovsky the news.

16. Jerry Adler et al., "Taking Offense: Is This the New Enlightenment on Campus or the New McCarthyism?" *Newsweek*, 24 December 1990, p. 48.
"Queer and Now" was written in 1991. Ken Wissoker thought up the title, and Mark Seltzer cheered me on with it.

QUEER

TUTELAGE

PRIVILEGE

OF UNKNOWING:

DIDEROT'S

THE NUN

Knowledge is not itself power, although
it is the magnetic field of power. Ignorance
and opacity collude or compete with it in mo-
bilizing the flows of energy, desire, goods,
meanings, persons. If M. Mitterrand knows English but Mr. Reagan lacks French, it is the urbane M. Mitterrand who must negotiate in an acquired tongue, the ignorant Mr. Reagan who may
dilate in his native one. Or in the interactive speech
model by which, as Sally McConnell-Ginet puts it,
"the standard . . . meaning can be thought of as what is
recognizable solely on the basis of interlocutors' mutual
knowledge of established practices of interpretation," it
is the interlocutor who has or pretends to have the *less*
broadly knowledgeable understanding of interpretive practice who will define the terms of the exchange. So, for instance, because "men, with superior extralinguistic resources
and privileged discourse positions, are often less likely to treat
perspectives different from their own as mutually available for
communication," their attitudes are "thus more likely to leave a
lasting imprint on the common semantic stock than women's."[1]
Such ignorance effects can be harnessed, licensed, and regulated on a mass scale for striking enforcements—perhaps especially around sexuality, in modern Western culture the most meaning-
intensive of human activities. The epistemological asymmetry of the
laws that govern rape, for instance, privileges at the same time men and

1. Sally McConnell-Ginet, "The Sexual (Re)Production of Meaning: A Discourse-Based
Theory," unpublished manuscript, pp. 387–88, quoted in Cheris Kramarae and Paula A.
Treichler, *A Feminist Dictionary* (Boston: Pandora Press, 1985), p. 264.

ignorance: inasmuch as it matters not at all what the raped woman perceives or wants just so long as the man raping her can claim not to have noticed (ignorance in which male sexuality receives careful education).[2] And the rape machinery that is organized by this epistemological privilege of unknowing in turn keeps disproportionately under discipline, of course, women's larger ambitions to take more control over the terms of our own circulation.[3] Or, again, in an ingenious and patiently instructive orchestration of ignorance, the U.S. Justice Department ruled in June 1986 that an employer may freely fire persons with AIDS exactly so long as the employer can claim to be ignorant of the medical fact, *quoted in the ruling,* that there is no known health danger in the workplace from the disease.[4] Again, it is clear in political context that the effect aimed at—in this case, it is hard to help feeling, aimed at with some care—is the ostentatious declaration, for the private sector, of an organized open season on gay men.[5]

Inarguably, there is a satisfaction in dwelling on the degree to which the power of our enemies over us is implicated, not in their command of knowledge, but precisely in their ignorance. The effect is a real one, but it carries dangers with it. The chief of these dangers is the scornful, fearful, or patheticizing reification of "ignorance"; it goes with the unexamined Enlightenment assumptions by which the labeling of a particular force as "ignorance" seems to place it unappealably in a demonized space on a never-quite-explicit ethical schema. (It is also, as I will be suggesting, dangerously close in structure to the more palpably sentimental privileging of ignorance as an originary, passive innocence.) The angles of view from which it can look as though a political fight is a fight against *ignorance* are invigorating and maybe revelatory, but they are dangerous places for dwelling. The writing of Foucault, Derrida, Thomas Kuhn, and Thomas Szasz, among others, has given contemporary readers a lot of practice in

2. Catherine A. MacKinnon makes this point more fully in "Feminism, Marxism, Method, and the State: An Agenda for Theory," *Signs* 7 (Spring 1982): 515–44.
3. Susan Brownmiller made the most forceful and influential presentation of this case in *Against Our Will: Men, Women, and Rape* (New York: Simon and Schuster, 1975).
4. Robert Pear, "Rights Laws Offer Only Limited Help on AIDS, U.S. Rules," *New York Times,* 23 June 1986, pp. 1, 13. That the ruling was calculated to offer, provoke, and legitimize harm and insult is clear from the language quoted in Pear's article: "A person," the ruling says for instance, "cannot be regarded as handicapped [and hence subject to federal protection] simply because others shun his company. Otherwise, a host of personal traits, from ill temper to poor personal hygiene, would constitute handicaps."
5. See also, in this connection, the discussion in *Epistemology of the Closet* (pp. 6–7) of the "complex drama of ignorance and knowledge" in the majority opinion of *Bowers v. Hardwick.*

questioning both the ethical/political disengagement and, beyond that, the ethical/political simplicity of the category of "knowledge," so that it seems, now, a naive writer who appeals too directly to the redemptive potential of simply upping the cognitive wattage on any question of power. The corollary problems still adhere to the category of "ignorance" as well, but so do some additional ones: there are psychological operations of shame, denial, projection around "ignorance" that make it an especially propulsive category in the individual reader, even as they give it a rhetorical potency that it would be hard for writers to forswear and foolhardy for them to embrace.

Rather than sacrifice the notion of "ignorance," then, I would be more interested at this point in trying, as we are getting used to trying with "knowledge," to pluralize and specify it. That is, I would like to be able to make use in sexual-political thinking of the deconstructive understanding that particular insights generate, are lined with, and at the same time are themselves structured by particular opacities. If *ignorance* is not—as it evidently is not—a single Manichaean, aboriginal maw of darkness from which the heroics of human cognition can occasionally wrestle facts, insights, freedoms, progress, perhaps there exists instead a plethora of *ignorances,* and we may begin to ask questions about the labor, erotics, and economics of their human production and distribution. Insofar as ignorance is ignorance *of* a knowledge—a knowledge that may itself, it goes without saying, be seen as either "true" or "false" under some other regime of truth—these ignorances, far from being pieces of the originary dark, are produced by and correspond to particular knowledges and circulate as part of particular regimes of truth. We should not assume that their doubletting with knowledges means, however, that they obey identical laws identically or follow the same circulatory paths at the same pace.

That a particular ignorance is a product of, implies, and itself structures and enforces a particular knowledge is easy to show, perhaps easiest of all, today, in the realm of sexuality. Because I want to use this novella itself, the eponymous character herself, as an exemplary figure for the economics of sexual ignorance, let me take examples of this axiom from the stormy editorial penumbra of Diderot's *La Religieuse,* an atmosphere agitated by the perennially interesting question of the innocence/ignorance of the virginal female. Naigeon, the 1798 editor of the novella that was widely published only after Diderot's death, expressed the fervent wish that he could have suppressed the book's lesbian scenes,

> the no doubt very faithful, but also quite disgusting depiction of the
> loathesome loves of the mother-superior. The various means she uses

to seduce, to corrupt a young child, whose ingenuousness and innocence she had every sacred reason to protect; this lively and animated description of her intoxication, her tumult, and the disorder of her senses under the very eyes of the object of her criminal passion; in a word, this hideous and true tableau of a kind of debauchery, quite rare moreover, but towards which mere curiosity could violently pull a mobile, simple and pure soul, can never be without danger for morality and health; and even when it does nothing but . . . hasten . . . that moment of orgasm marked by nature, when the desire, the general and common need to take pleasure and multiply, hurls one sex with furor onto the other, even that would be a great evil. I often made this observation to Diderot; and I should say here, to exonerate the philosopher in this respect, that, struck by the reasons on which I based my opinion, he was quite determined to make to decency, shame, and the moral conventions this sacrifice of a few pages that would be cold, insignificant, and boring to a man, even the most dissolute, and revolting or unintelligible to a decent woman. Certainly, purged in this way, the work would not have lost anything of its effect. . . .

Today, these omissions . . . would be entirely useless. The first impression [*la première impression;* also, the first printing], always so difficult to efface, is made; and all the art, all the talent of Diderot, applied to the correction, to the perfecting of these two stories, could neither destroy nor even weaken it in the minds of most readers. Some from that strange mania for having all the works, without exception, of a philosopher, a poet, or an illustrious writer; others on a whim or from envy, and from that more or less keen need of all mediocre men to console themselves for their nullity by depreciating the greatest geniuses and eagerly seeking out their faults, stubbornly persisted in demanding again *La Religieuse* and *Jacques le fataliste* just as they had initially been published.[6]

The unknowing female reader on whose simple and pure soul the novella's scenes of lesbian desire would exert at the same time a repulsion, an effect of unintelligibility, and a violent attraction is a creature into whose adaptable construction as much care must obviously have gone as into the male reader who manically persists in his demands that a book include passages that are cold, insignificant, and boring to him. Assézat, the 1875 editor who

6. Quoted in J. Assézat, *Oeuvres complètes de Diderot* (Paris: Garnier Frères, 1875), vol. 5 (*Belles-Lettres* II), pp. 209–10. Cited in the text as JA.

quotes these strictures, asks with good reason, "isn't one tempted to see in his scruples nothing more than the revenge of an editor beaten to the punch?" (JA, p. 210)—insofar as the ineffaceable first impression, some-how always already stamped, by which the blankness of the female reader is rendered indistinguishable from her terrifying appetitiveness, seems to be identical to the unfortunately definitive and demand-producing first print-ing of *La Religieuse*.

At the same time, however, the rule that *it takes one to know one* seems to operate for editorial *ressentiment* as strongly as for prurient sentimen-tality. For it is Assézat, the later editor, who writes:

> We call *La Religieuse* a masterpiece, and it is such a masterpiece that it cannot be touched without losing part of its value and without be-coming, even, dangerous.* . . . To the ignorant, [Diderot] teaches noth-ing; to one who knows, he is very far from telling all.
>
> On this particular point, Naigeon has spoken idiocy, and it hardly befits the man who added the chapters we have marked in *Les Bijoux indiscrets* to cross himself hypocritically before a page, a single one, which one cannot reproach except with having understated the truth.

---

* This is what happened with M. Génin's edition of *La Religieuse* . . . . The ellipses that replaced certain passages, those mysterious ellipses, seemed huge with horrors and monstrosities, and, certainly, made the young people dream more than the text itself would have done. These inept reticences are like the thoughtless questions of confes-sors. (JA, pp. 4–5)

Here again, it is only a very pointed and well-drilled readerly ignorance that can be at the same time so obdurate as to be taught nothing by the graphic Diderot, and so suggestible as to be inflamed by a row of dots; and the active blue pencil of the imagination that had let Naigeon see young women as eternally corrupted by that first impression is nicely answered by Assézat's branding demarcation at once of the shame-inducing (speci-fied) additions to *Les Bijoux indiscrets* and of the totemic, or taboo, single page of *La Religieuse* that cannot be be so much as touched—by an edi-tor—without extreme sexual danger.

If the inveterate interlinearity of knowledges with their own ignorances is one of the lessons of all this editorial swaddling, however—as though, like George Eliot characters, incisive or susceptible texts can't go out into the world until they have been well-wadded with stupidity—the more stir-ring fact is with what officious efficacy, with what a long and muscular arm the very specific unknowing *in* the novella has mobilized the punitive energies of these two readers. And not only these readers: the illusionistic force of Suzanne Simonin's narrative depends, I'd argue, on its success in pulling any given reader into a projective, protective and punitive relation to a tellingly deployed sexual ignorance.

Part of the interest of *La Religieuse* for a historical exploration of gay-related discourse is that it comes from a time before the rich and murderously contradictory modern array of ways of thinking *gay identity,* especially in relation to gender identity, had begun to emerge. Both because of the mid-eighteenth-century origin of the novella and because of its conventual venue, the question of lesbian sexual desire—*is* what is happening sexual desire, and will it be recognized and named as such?—looks, there, less like the question of The Lesbian than like the question of sexuality *tout court.*[7] It is curious from the late twentieth century to come again under the sway of a text and a narrating voice that emerge from that brief, and itself fairly murderous, interregnum or overlap between the rule of the priest and that of the doctor. Assézat, the 1875 editor, writing firmly from the age of the doctor, defends Diderot's sexual graphicness—or rather, seems to show it to be inevitable—in these terms:

> How could it be wished that Diderot should halt in his path? What was it he wanted to depict? The life of the cloisters. And should he have left out one of the forms of hysterical illness which results from them most often, not to say always? Cruelties one may deny; they occur behind closed doors and only rarely become known . . .; but illness speaks, and always loudly, and it begs for the intervention of a man, who is no longer the priest, but the doctor. (JA, pp. 4–5)

"La maladie parle, et toujours haut, et elle réclame l'intervention d'un homme." Call it the voice of illness or call it otherwise; the voice that emerges from Diderot's novella does possess to an uncanny degree the penetrating insistence of demand, and demand *of men,* that Assézat invokes. Marie-Suzanne Simonin, from the moment one stops hearing her as the transparent wisp of impossible passivity that would provide the merest blank pretext for sentimental narrative, is unmistakably a pain in the neck. "Certainly I surpassed my sisters in qualities of mind and beauty, character and ability, and this seemed to upset my parents."[8] You can take that

---

7. Those who read it otherwise have not, however, been lacking; see, for example, Vivienne Mylne, "What Suzanne Knew: Lesbianism and *La Religieuse,*" *Studies on Voltaire and the Eighteenth Century* 208 (1982): 167–73.

8. Denis Diderot, *The Nun,* trans. Leonard Hancock (Harmondsworth, Middlesex: Penguin Books, 1974), pp. 21–22. Further citations from *The Nun* in this essay refer to this edition.

as the beginning of a fairy tale, and hence not quite take it, or else you can take it as your introduction to a young woman, charismatic, deaf to herself and dangerous to her intimates, at the same time profoundly oppressed, in whose unmodulated and rather afflicting presence real time will be spent. A twentieth-century reader might think to describe the effect by calling Suzanne a real neurotic, but again the medically pseudo-specified neurosis provides hardly more than an opaque backing for the spectacular adjective "real." The shocking illusion of reality depends, grain for grain or volt for volt, on the skepticism, resentment, and anxiety that the character is able to mobilize in the reader.

If one tries to read *La Religieuse* as the epistolary testimony of a conventional heroine—as the letter, that is to say, of a tabula rasa, of a passive and denatured innocence that is the object of "ignorant, warped, and unnatural" religious "abuses" which it is the story's project to denounce[9]—one reads an insipid and inconsistent book whose continually unraveling illusion is not at best an excessively interesting one. The game of confronting Suzanne's (angelic and natural) *innocence* with the Church's (demonic and unnatural) *ignorance* requires a good deal of editorial encouragement (from Naigeon, Assézat, Tancock . . .) to be kept afoot. The crowning improbability is the one remarked on by the translator, Leonard Tancock, who does nevertheless read the book in this way. The heroine, he says, because of "the psychological necessities of the novel," "must be"—note the oddly unanchored imperative of her ignorance here—

> must be young and impressionable throughout her story. . . . In order to give sustained interest Suzanne has to be subjected to a series of trials of increasing severity. But the very fact that she must be afflicted afresh with new and more acute fear and anguish means that she must remain throughout the innocent victim in spite of the manifest presence around her of evil and perversion, and this with her obvious intelligence and in a tale told retrospectively by a narrator perfectly aware of the end. . . . She is wonderfully observant, with Diderot's eye for the characteristic gesture or facial expression, but in spite of all her experience, including having been accused [and, one might add, accusing others—see p. 57 of Tancock's edition] of masturbation or homosexual practices, she is completely innocent and uncomprehending in the face of the homosexual approaches of the Superior of Sainte-Eutrope. It is true that the evolution of the Superior from a kindly . . . person to a sinister Lesbian is beautifully graded, but it is stretching

9. Quoted from Leonard Tancock's introduction, Diderot, *The Nun*, p. 11.

credibility to [the] breaking point to make Suzanne quite unaware of the meaning of her Superior's behaviour when it comes to describing an orgasm in almost clinical detail. (p. 13)

Perhaps especially so when it's Suzanne having the orgasm (pp. 138, 142–43)? According to Tancock's way of charting the passive sufferings of Suzanne, her story is meant to illustrate the truth that "power corrupts, especially the power of women in authority over other women"; it demonstrates "four kinds of danger in the segregated life, and these correspond to the successive experiences of Suzanne Simonin. They are madness, the paralysing effect of a saintly, mystical personality who inspires blind hero-worship, sadistic cruelty and bullying, and homosexuality, which can include elements of the other three" (pp. 15–16).

Tancock, in short, in order to avoid attributing any shred of power, volition, or knowledge to Suzanne—in order to preserve her as a space of idealized ignorance and passivity—finds it necessary to impugn her integrity as an artistic creation (he considers charging Diderot with "simple negligence," with "*naiveté*," or with being "frankly pornographic," finally letting him plea-bargain to the more amiable misdemeanor of "losing himself in what he is doing at the moment" [pp. 13–14]); and he is required as well to resort to the reproduction of women-hating and gay-hating *idées reçues* that are manifestly at odds with the supposedly modern "sympathy," "understanding," "common sense," and "tolerance" with which he claims to find Diderot treating sexual variety (p. 16).

In an almost opposite, and vastly more interesting and respectful, reading of Suzanne's epistemological situation, Rita Goldberg in *Sex and Enlightenment: Women in Richardson and Diderot* sees Suzanne not as the personification of a passivity untouched by knowledge, but instead as the inaugural figure of a historical process that she calls "the feminization of knowing." [10] Without at all denying Suzanne's victimization, she finds more arresting the relation of that to her peculiarly active will toward knowledge, which she sees condensed in Suzanne's cool gaze. "It is rather as if the corpse on the dissecting table were also the doctor doing the dissection. . . . Rather like a Cubist nude, Suzanne is able to look upon her own body and out upon the world, even if the planes of her own existence as a character are thereby fatally realigned" (RG, pp. 186–87). Furthermore, although Goldberg sees Suzanne as entirely passive except for the gaze that repre-

---

10. Rita Goldberg, *Sex and Enlightenment: Women in Richardson and Diderot* (Cambridge: Cambridge University Press, 1984), p. 201. Further cited in the text as RG.

sents her sheer force of consciousness, that exception is a most consequential and indeed active one, its power measurable by its ability to destroy.

> Her very presence seems to be enough to make her three Superiors aware of their own failures. Each of them is harmed, even destroyed, by her steady gaze. . . . Suzanne causes upheavals, chemical changes, as it were, in the objects of her gaze. Like Diderot's interpreter of nature in the scientific works, Suzanne forces her environment into giving up its secrets. The process of knowing is apparently violent. (RG, pp. 180–81)

This understanding of Suzanne as "an embattled philosopher, fighting for the right to know and to act freely in a world where the exercising of these faculties is by no means yet perceived as God-given" (p. 203), however, although it does justice to the vibrant willfulness of Suzanne's effects on her world, nonetheless by describing her as *pure knowing* (or, in Goldberg's term, "epistemological transparency," p. 194), reproduces some of the problems that Tancock and his predecessors had encountered in trying to preserve her as a space of *pure unknowing*. Specifically, there is once again a problem about Suzanne's obstinate sexual incomprehension; and once again, the only move from there seems to be a gesture of disavowing Suzanne as the locus of an illusion of specific individual identity. "Suzanne's lack of awareness in sexual matters is," Goldberg says, making a point of it,

> in itself not particularly important. One could think that, like the high-pitched transvestite ladies of pantomime, Diderot may have simply got the tone wrong. . . . If one wanted to argue for the accuracy of Diderot's portrayal of character, one could always point to Suzanne's loveless background, with which this inability to love in adult life would be consistent.
>
> But Suzanne is not a "character" in this sense. She has no history and no future. (pp. 183–84)

This character who is not a character easily remains in possession of her peculiar "epistemological" "purity" (p. 193); her "innocence" finally remains through this reading "particularly unsullied" (p. 185), and however potent her gaze may be in its effect on the institutions that imprison her, it is with a potency untinged by the shadow of any enfoldment, any psychological opacity or tendentiousness or difference, of sheer personal demand or disposition. This critical complex of epistemological purity and psychological evacuation takes its worst toll on the range of *readerly* response that it makes permissible to talk about: after a description of Suzanne that is far slyer and more nuanced than I have been able to indicate here, Goldberg is reduced to a description of herself in her relation *to* Suzanne that is by comparison almost abjectly flat. "Nonetheless, Suzanne does succeed in moving us. . . . We must admit that Suzanne can bring tears to our eyes, as she did to those of her creator." But Goldberg has also made explicit the specifically punitive relation to Suzanne that is being atoned for by this uncharacteristic ascesis of language and intelligence: "*Perhaps the criticism in these pages appears unfairly harsh. It is not meant to be. After all our reservations and mockeries have been disposed of,* we must admit that Suzanne can bring . . ." (p. 202, emphasis added).

The reader who ceases to make a fetish out of Suzanne's inviolable passivity or the immaculation of her experiential surface finds, on the other hand, both a livelier and an oddly more coherent story inscribed on the same pages. When a woman writes, "It was ordained that in that establishment *I* should suffer all the ills it is possible to go through" (p. 114, emphasis added), as a way of introducing the lingering death—from an illness caught while nursing herself—of the only other nun there who has stood by her, shouldn't it be possible to see that woman as a stubbornly selfish one? She had earlier, without consulting this other nun, plunged her into mortal danger by thrusting a wad of forbidden papers at her: "Risking anything that might happen *to me* I resolved to entrust my document to her" (p. 61; emphasis added). In fact, usage after usage of the first person in this first-person narrative has the heartily alienating effect of incipient italics: "with my natural candour" (p. 21) she laments, "Oh, how many times I wept because I was not born ugly, foolish, silly, conceited . . . !" (p. 22) Of her first Mother Superior after she is forced to take her vows she reflects,

> If she made any distinction it was based on merit, and that being said I don't know whether I ought to tell you that she had a great affection for me and that I was by no means the least of her favourites. . . . If there were anything I could find fault with in Madame de Moni it would be that she let her taste for virtue, piety, candour, meekness, talent and integrity be seen too obviously . . . . They were unfortunate indeed whose confidence she had difficulty in winning! They must be bad, irretrievably bad, and know it. (pp. 46–47)

Meekness indeed. Suzanne never leaves the labor of her praise to others: "I have a striking face," she confides; "I have a voice which touches the heart and my expression bears the stamp of truthfulness" (p. 94). As indeed it does.

Psychologically, this vulgar heft of egotistic annunciation makes good sense in the survivor of the childhood that Suzanne describes. She creates through sheer assertion a space in the world for a woman for whom there is no space in the world. One of three daughters in a bourgeois family (M. Simonin is an advocate), she is however the child of her mother's adulterous liaison with a man now dead; M. Simonin views her with repugnance and her mother with dread. "I must confess," her mother volunteers with the apparently matrilineal candor, " 'you are a constant reminder of such hateful betrayal and ingratitude on the part of another man that I cannot bear the thought; the vision of this man always rises up between you and me, he spurns me, and the loathing I have for him recoils upon you. . . . My child, don't poison my life any longer.' " (p. 40). A certain florid expressiveness in this language of maternal refusal is, moreover, the closest thing to warmth that Suzanne is ever shown by any member of her family; far more characteristic of them is the simple desire that, though no one wants to make the expenditure of energy involved in doing away with her, Suzanne *not exist*. Certainly she must not be visible, for instance at the tableau, foreseen by the mother, of the mother's death:

> "For that must come, and at that terrible moment your sisters will be at my bedside—how can I see you with them, what effect would your presence have in these last moments? My daughter—for such you are whatever I do—your sisters have by law a name you only have because of a crime—don't [by your presence] torture your dying mother . . . ." (p. 41)

The narrowness, the virtual nonexistence, of the space for Suzanne's being is made excruciatingly clear, less even in Suzanne's suffering than in her mother's enforced parsimony toward her. Her mother is not a loving woman toward Suzanne, but even if she were, she could not entitle her daughter to any ontological ground broader than the already eclipsed intersection between the female-punishing structure of her proscribed sexuality (the adultery) and the female-punishing structure of her economic life (the marriage). Whatever the mother's intentions or desires, Suzanne could still hope only to occupy the barest constricted margin of a bourgeois excess. Her heritage is less the cheese paring, even, than the sweat of the cheese paring. "I hope I shall have nothing on my conscience when I die," her mother explains:

> "—I shall have paid for your [convent] dowry out of my savings. I don't take advantage of my husband's open-handedness, but I put

aside day by day what he gives me now and again out of generosity. I have sold such jewels as I had, with his permission to spend the proceeds how I like. I used to like gambling, but I never do now; I used to love the theatre, but I have done without; I used to love society, and I live in retirement; I used to like display, but I have given it up. If you take the vow, which is my desire and Monsieur Simonin's, your dowry will be made up out of what I manage to put by each day." (pp. 40–41)

The moral as well as monetary tightness of fit in this female economy becomes even more explicit in the mother's actual deathbed testament, when she sends Suzanne "what I have managed to save out of little gifts from Monsieur Simonin" (p. 53):

"Pray for me; your birth is the one serious sin I have committed. Help me to expiate it. . . . God, who sees all, will in His justice mete out to me all the good and all the evil you do. Farewell, Suzanne, ask nothing of your sisters, they are not in a position to help you; hope for nothing from your father, he has preceded me. . . . Farewell once again. Oh, wretched mother and unhappy child! Your sisters have come, and I am not pleased with them; they are taking things, carrying things away, and in front of a dying mother indulge in greedy quarrels which are torturing me. When they come near my bed I turn away, for what would I see? Two creatures in whom lack of money has killed all natural feeling. . . . They have had suspicions, I don't know how, that I might have some money hidden between my mattresses, and there is nothing they have not had recourse to in order to make me get out of bed. And they succeeded, but fortunately the person to whom I am entrusting this money had come the day before, and I had given him this little package, with this letter which he has written to my dictation. Burn this letter. . . ." (p. 54)

The temporal space in which this letter could have been written and the money sent is, of course, a self-contradictory and hence impossible one; the desperate plea to "burn this letter" cannot be necessary since the time of its writing and sending has already been pared away at both ends so finely as to leave nothing whatever in the middle. Nonetheless, the perseverating desperation of the selfish female voice survives—survives also the preemption by daughters (whose relentlessness is only another fleshing-out of the very same lack) of the moment of the mother's death; survives later in the hating perversity, indistinguishable from a love of life, of a daughter who will refuse to commit suicide for the sole reason that she believes that other women (her Mother [Superior], her Sisters) wish her to do so (p. 59).

Subordinated to her inferior sisters and banished inexorably to a conventual life "so that the mother might atone for her sin through her child" (p. 40), Suzanne in the letter that comprises most of the novella tells the story of her succession of convents and Mothers Superior and of her succession of struggles to get free of her vows, and loudly renews her claim for male rescue. The legalistic rationality of this claim—making her seem indeed the "embattled philosopher" Goldberg describes—lets her show her skill in manipulating the bare overlap of contradictory ideologies: "vocation," for instance, can be used in her language either as a priest would use it or as the vocational counselor in a public high school would; and a certain inconspicuous elasticity in her notion of *consent*, around the taking of vows, is harmonious with the fact that on one occasion she had dramatically refused these vows but on another had repeated them without any protest at all, but in, she says later, a state of "physical absence" (p. 52) or psychological unfreedom (p. 75). From her stranglingly constricted space in the knotting together of a market economy with a theological one, each moreover differently inimical to women, Suzanne has acquired both a cunning and a persistence in playing one off against the other in the small, treacherous vocabulary common to both. Where Suzanne is insufficiently cunning, however, is in her fantasy that there is a place *elsewhere* in which her desire to survive could operate differently: a place called, in the vocabulary she here recklessly borrows from one side only of her double bind, "freedom." The brief, fragmented narrative of the events that follow her escape—an apparent rape, confinement in a brothel, squalid poverty, grinding work, physical injury, and the utter attenuation even of her ability to narrate, her will, her consciousness ("I tremble like a leaf, I cannot hold myself up and my work drops out of my hands. Almost all my nights are sleepless, and if I do sleep it is only fitful dozing, and I talk, cry out, shout, and cannot imagine how people round me have not yet guessed my secret" [pp. 186–87])—demonstrates that, whatever margin of energy and rhetorical leverage the idea of "freedom" may afford her in the confinements of her family and her convents, "freedom" is not after all a place and its location is not the outside of enclosed institutions.

We have so far described Suzanne as most active "toward" freedom, toward an outside space. Even within and in relation to the convents, however, although Suzanne may be oppressed or even tortured, she is never more than momentarily passive. In fact, she is especially insistent about her agency in the one, prolonged instance where she is the victim of horrifying conventual sadism and torture. On the death of her first, sainted Mother Superior,

> I hastened to make my position worse. . . . The first thing I did was to give myself up to all the grief I felt at the loss of our first Superior, and sing her praises on every occasion, make comparisons between her and our present ruler which were unfortunate for the latter. . . . I read and re-read our regulations and knew them by heart. . . .
>
> . . . I had let some indiscreet remarks escape me on the subject of the suspicious intimacy between some of the favourites; moreover the Superior went in for long and frequent interviews with a young priest, and I had discovered the reason and the pretext for these. I omitted none of the things needed to make myself feared and hated and to bring about my ruin, and in this I succeeded. (pp. 55–57)

True to her Diderotian origin, Suzanne is a ham, forever creating in these convents real-life tableaux and *coups de théâtre* to feature herself, and speculating about or frankly describing their effects on her audience (e.g., pp. 31–32, 33, 45, 59, 71, 112, 153). She is also, however, a born leader and a born scapegoat. She keeps each of the convent regimes—three are described in detail—whipped up to a high lather, whether in celebration, envy, or punishment of her.

Suzanne's various mobilizations of these establishments do, nevertheless, have a certain consistency of result. Wherever she goes, she creates, from the Mother and Sisters available, a new family that centers on herself—sometimes one that will nurture her, sometimes one that threatens her eradication. But once each of these Suzanne-centered families has been established, a similar narrative must be played out: one in which Sisters are displaced and the Mother is publicly discredited. The first Superior, the spiritually powerful Madame de Moni who had loved her so deeply, somehow through her intercourse with Suzanne is bereft of her charismatic gift, abandoned by the male deity: "I don't know what had been happening inside her, whether I had filled her with misgivings about her own strength which never left her again, whether I had made her self-conscious, or whether I had really broken her communication with God, but she never regained her gift of consolation" (p. 48). The Mother Superior who replaces Madame de Moni is the one with whom Suzanne, willfully by her own account, stirs up a drama of blood-curdling sadistic enactment, with herself as scapegoat and the other Sisters as either incidental sacrifices (like Sister Ste.-Ursule) or auxiliary tormentors. Suzanne succeeds in exposing this "weird" (p. 81) and abusive regime to the men who represent ecclesiastical authority, although not in achieving her own release from her vows; she is merely moved to a new convent. There, at Arpajon, she becomes the object of an amorous seduction by her third Mother Superior, displacing the previous favorites. When Suzanne reports her Superior's activities to her confessors, they instruct her to spurn her advances violently, and the Superior dies in protracted scenes of agonizing madness and in disgrace.

It is Suzanne's relation to this third Superior that has been the occasion for so much editorial bustle, for it is here that the question of sexual motivation—implicit at most in the relations with the first two Superiors—comes inescapably to the surface, and it is also here that Suzanne's role in the annihilating ruin of the Mother is dramatized most legibly. Here, too, the epistemological questions around Suzanne become critical. Reporting in fine-grained detail on the Mother Superior's every gesture, request, touch, complexion engorgement, gasp, and swoon—and reporting, for that matter, in equally high resolution on her own—Suzanne nonetheless insists both to the Superior and to the reader that she is entirely at a loss about what can possibly be going on.

> By now she had raised her collar and put one of my hands on her bosom. She fell silent, and so did I. She seemed to be experiencing the most exquisite pleasure. She invited me to kiss her forehead, cheeks, eyes and mouth, and I obeyed. I don't think there was any harm in that, but her pleasure increased, and as I was only too glad to add to her happiness in any innocent way, I kissed her again on forehead, cheeks, eyes and lips. The hand she had rested on my knee wandered all over my clothing from my feet to my girdle, pressing here and there, and she gasped as she urged me in a strange, low voice to redouble my caresses, which I did. Eventually a moment came, whether of pleasure or of pain I cannot say, when she went as pale as death, closed her eyes, and her whole body tautened violently, her lips were first pressed together and moistened with a sort of foam, then they parted and she seemed to expire with a deep sigh. . . . I looked at her, wild-eyed and uncertain whether I should stay or go. She opened her eyes again; she had lost her power of speech, and made signs that I should come back and sit on her lap again. I don't know what was going on inside me, I was afraid, my heart was thumping and I breathed with difficulty, I was upset, oppressed, shocked and frightened, my strength seemed to have left me and I was about to swoon. And yet I cannot say it was pain I was feeling. . . . We remained in that peculiar state for some time. (pp. 137–38)

It is not too surprising that editors "worry" about whether such epistemological effects may be pornographic in intent—their aptitude to charm, at any rate, is undeniable; nor is it surprising that the high contrast in Suzanne of descriptive exactitude with diagnostic obtuseness should occasion incoherence in any reading that is based on her "epistemological purity," whether a purity of unknowing or of knowing. For that matter, the mixed desire and incoherence it provokes in such a reader is not so different from its effect on the Mother Superior.

If, however, one were to ask about Suzanne's very pointed ignorance here, not what epistemological essence it attaches to her as if allegorically, but more simply *what it makes happen,* there are readier answers to be had. To begin with, it is, as a score of citations could testify, itself seductive. Suzanne's ignorance permits her, moreover, to participate in all the physical pleasures without any of the guilt or anxiety of her Mother Superior's sexuality—"enjoying ourselves," as she suavely puts it at one moment when the woman has expired between her breasts, "in a manner as simple as it was pleasant" (p. 135). Resistant as she is to associating these scenes with sexual guilt, she is even slower to associate them with sexual jealousy; so that she is able to inflame literally to madness the jealousy of the Sister she has supplanted in the Mother's love, without having to see her usurpation for what it is. When the Sister asks her to try not to see the Superior so often,

> This seemed to me such a strange request that I couldn't help replying: "How does it affect you whether I see our Superior often or not? I don't mind in the least if you see her all the time. You shouldn't mind any the more if I do the same. Isn't it enough for me to protest that I shall never damage you or anybody else in her eyes?"
> She made no reply except in these agonized words as she tore herself away from me and hurled herself on to her bed: "I am a lost soul!" (pp. 131–32)

Suzanne's unknowing makes her invincibly attractive to the Mother and invincibly triumphant over the Sisters, even while—that is, exactly because—it insulates her from any consciousness that the most volatile of power negotiations is being conducted around and by herself. Is it any wonder that, far from passively *lacking* the knowledge that what is going on constitutes "sexuality," she actively and even lustily *repels* it? She protests to the Superior,

> "I know nothing, and I would rather know nothing than acquire knowledge which might make me unhappier than I am now. I have no desires, and I don't want to discover any I couldn't satisfy."

"But why couldn't you?"

"How could I?"

"As I do."

"As you do! But there is nobody in this place."

"I am here, my dear, and so are you."

"Well, what am I to you, and you to me?"

"How innocent she is!"

"Oh yes, I am, it is true, dear Mother, and I would rather die than cease to be so." (pp. 145–46)

Suzanne is not the only person committed to maintaining this powerful but labor-intensive ignorance. Her male confessors, eager as they are to know from her everything that goes on in the convent, are nonetheless as keen as she is for its preservation; one of them says to her about the other,

"abide by his advice, and try as long as you live to remain ignorant of the reason."

"But it seems to me that if I knew what the peril was I should be all the more careful to avoid it."

"But perhaps it would work the other way."

"You must have a very poor opinion of me."

"I have the opinion I ought to have about your morals and innocence, but believe me there is such a thing as poisonous knowledge which you couldn't acquire without loss. . . ."

"I don't understand."

"All the better." (pp. 175–76)

Suzanne's ignorance is precious to the men of the church hierarchy, not because it insulates her from sensations of which they disapprove—as we have seen, she partakes of those sensations whether or not she names them —but because it allows her to be such a tympanically responsive transmitter of the convent's doings to them and of the paternal law back to the convent.

"Without daring to explain myself in any more detail for fear of myself becoming an accomplice of your unworthy Superior and, with the poisonous words that would come from my mouth in spite of myself, withering a delicate flower that is only kept fresh and stainless until your age by the special protection of Providence, I order you to avoid your Superior . . . . If this wretched woman questions you, tell her everything and repeat what I have just said, tell her it were better for her that she had never been born or that she had thrown herself into hell by a violent death." (p. 162)

Suzanne, her mediumistic opacity (or transparency?) overriding her normally friendly disposition, carries out with a customary buoyancy the orders of her confessors and hounds the Superior to a terrible death.

The boundary of the power conferred by Suzanne's ignorance is in principle that she has only the power *of being manipulated,* manipulate that power how she may. We do see that she carefully keeps herself in a position to choose by whom she will be manipulated, though. The Superior, attracted though she is to Suzanne's unprecocity, keeps wanting to verbalize "the language of the senses" (p. 145) for her: to try to stabilize the relationship, to compromise Suzanne by forcing her to see herself *as* compromised by her pleasure, to try to make her a bit less of a loose cannon. Suzanne will have none of it. But although she refuses to be taught that the scenes between herself and the Superior have verbalizable content, she refuses also to be content *not* to verbalize them. She hovers seemingly forever about the threshold of broaching them at confession (e.g., p. 150). The Superior, unable to obtain Suzanne's verbalized complicity, next tries to help her think of their encounters as something beneath verbalization, like the unproblematical encounters in Carroll Smith-Rosenberg's "Female World of Love and Ritual";[11] but in this, too, she fails.

> "I order you not to talk to [your confessor] about such nonsense. It is pointless for you to go to confession if you only have silly little things like that to tell him. . . . [W]hat wrong do you think there is in keeping quiet about what it was not wrong to do?"
> "Then what is wrong with mentioning it?" (pp. 160–61)

The situation that Suzanne through her ignorance finally manages to create is one in which there is, as there has been from the start, a question of her own legitimacy, formulated—this time—as the legitimacy of her desires and sensations. But this time there are two actively competing authorities, that of the Mother and that of the Father, over the question of legitimacy; and *it is at last up to her to legitimate one or the other of them as the legitimator of female desire.* One night when they are close to curling up in bed together, the Mother says,

> "The whole place is asleep around us, and nobody will know anything. I am the one who rewards and punishes here, and whatever your confessor may say I cannot see what harm there is for one friend to take

11. Carroll Smith-Rosenberg, "The Female World of Love and Ritual: Relations Between Women in Nineteenth-Century America," *Disorderly Conduct: Visions of Gender in Victorian America* (New York: Oxford University Press, 1985), pp. 53–76.

into her bed another friend who is very distressed . . . . And am I not your dear Mother?"

"Yes you are. But it is forbidden."

"My dear, I forbid it in others, but I allow it to you, and am asking you to do so." (p. 150)

It is, however, impossible to Suzanne—though she is finally in a position to do so—to underwrite the Mother's authority to desire and to confer permission to desire. One of the last scenes between Suzanne and the Superior, well on her way to insanity and death, is an especially vengeful and cruel inversion of what had been a terrifyingly cruel scene between Suzanne and her real mother.

> She was my Superior and here she was at my feet, her head was on my knee which she was grasping. I held out my hands and she eagerly clutched them, kissed them and looked at me again. I raised her to her feet. She staggered and could hardly walk, and so I led her back to her cell. When her door was open she took me by the hand and gently tried to pull me in. . . .
>
> "No," I said, "dear Mother, no. I have made a vow, and it is best for you and me. . . ."
>
> "You won't, Sainte-Suzanne? You don't know what the result may be, no, indeed you don't. It will kill me."
>
> These words produced exactly the opposite effect on me to the one she was aiming at, for I snatched my hand away and fled. (p. 169)

Suzanne's repulsion and terror at the (to her unacknowledgeable) sexual need of the Mother abject at her feet may be the only possible answer to the unanswered appeal she had made to her own mother in the earliest pages of the story, before she had formed any idea about the secret of her mother's sexuality and her own birth:

> I flung myself at her feet and laid my head on her knees, saying nothing, but sobbing and gasping for breath. She pushed me away roughly. I did not get up, my nose began to bleed, I nevertheless seized one of her hands and made it wet with mingled tears and blood as I kissed it, saying: "You are still my mother and I am still your child." She pushed me even more roughly, snatched her hand away from mine and said: "Get up, you miserable girl, get up." I obeyed, sat back on the seat and drew my hood over my face. She had put such firm authority into her tone of voice that I felt I ought to spare her the very sight of me. My tears and the blood from my nose mingled together and ran down

my arms, and I had it all over me before I noticed. I gathered from the few words she said that her dress and underclothes had been stained and that it annoyed her. . . . On the way up the stairs I once again flung myself at her feet and held her by her dress, but the only result was that she turned round and looked at me, and there was such indignation in the set of her head, her mouth and her eyes that you can imagine it better than I can describe. (p. 35)

To the need of the Mother who appeals to her from the same position, Suzanne can only bring a vindictively echoing numbness. Instead of responding to the Mother, she reports to, and follows the orders of, the young male confessor who in the end rescues her from the place of women (the place where "there is nobody") only apparently to rape her and abandon her to the abattoir of the also male-dominated free market.

I am not blind, any more than Rita Goldberg is when she suppresses her "unfairly harsh" "mockeries" of Suzanne, to the political as well as tonal dangers of emphasizing Suzanne's active deployment of epistemological power, and with it the extreme partiality of the uses she makes of it. The tradition of attributing power to the resistant object of sexual siege rather than to the besieger who is abusing a more palpable authority is too long and destructive for me to wish to add to it. I fear: the sexual harasser in the workplace or classroom will defend himself with too similar reflections on the pointedness, hence the implausibility, of the "innocence" or "ignorance" of the person whose life he is making miserable; and another too similar train of reflection in psychoanalysis/psychiatry has contributed for decades to the discrediting—"She really wanted it; she was being seductive; it was a hysterical fantasy"—of the voices of children and former children brave enough to bear witness to sexual abuse. Walter E. Rex, in his study of *La Religieuse,* makes this punitive case precisely against Suzanne in the crudest possible terms, revealingly: "Suzanne gets exactly what she had been asking for." [12]

12. Walter E. Rex, "Secrets from Suzanne: The Tangled Motives of *La Religieuse,*" *The Eighteenth Century: Theory and Interpretation* 24 (Fall 1983): 193. The viciousness against Suzanne that Rex betrays in a basically sympathetic essay whose overarching intention appears to be to validate the possibility of a lesbian choice for her (pp. 197–98) is a striking artifact of the compulsive symmetry he imposes on the alternatives (of knowledge, of power) that he presents. For instance, "the creative forces [in Diderot] that brought into being this strangely powerful indictment of the cloister, also—through an equally strange reversed pull of the opposite—brought into being an anti-indictment, in fact, an antidote. . . . Although in the convent all are victims, since they are controlled, confined, and punished by that society's

A simple response to this worry would be that the deglobalizing, the pluralizing of the idea of innocence/ignorance, that we are trying to effect here, should also if it were consistently understood result in a disassembly of the fascination with reified innocence/ignorance *as* the main ground for claiming injury, especially sexual injury. *For instance,* there is no information about one's previous history that should be relevant to the evaluation of one's claim to be a victim of rape; *for instance,* it should not require a finding that Suzanne was entirely without her own knowledge, her own weapons, her own interested motives, to authorize a resistance that would under that finding be either impossible (from her) or dangerously vicarious (for her). One need not be an implausible, hence always a discreditable, tabula rasa, a heroine of passivity, an incarnation of unknowing or of transcendent knowing, in order to register injury including sexual injury. In fact, it might well strengthen our ability to define and hence to combat such injury if we could understand it in some way that separated it *entirely* from the issue of initiation, of the violation of a supposed originary non-knowledge, non-desire, non-individuation that can perhaps be shown in every case already to have been breached.

I think, then, that a reading like the one performed here might be shown not *necessarily* to involve an impulse toward the discreditation or punishment of Suzanne. A sufficiently thoughtful and rationalized boundary between the demands of the reader and the needs and strategies of the character can be instructive: if readers can give up the sentimental requirement of finding a unitary epistemological field in the heroine, then Suzanne's politically telling, if finally unsuccessful, deployment of multiple kinds of knowledge and an associated plurality of ignorances can be read and appreciated without risking yet another brush stroke added to the great historical mural called "Blaming the Victim."

On the other hand, to render such a reading innocuous would require that the relation of reader to character be patrolled with a sanitary exactitude that must a good deal attenuate its force. It is possible that there is more to be learned by attesting than by dismantling the anger that a character like Suzanne arouses. In fact, my own experience is, as I've suggested,

---

institutions, yet—to turn the coin over on the other side—all can also be viewed more actively as forming the convent, creating its life, applying its rules, and hence, responsible for (and guilty of) its punishments" (p. 193). This gender-blind, power-blind compulsion toward symmetry requires that as soon as Suzanne exercises any power or choice at all, then everything that happens to her, however brutal, must be considered not partially, not contingently, but *exactly* "what she had been asking for."

that Suzanne persists and coheres as a character *first* in the medium of that anger against "her" and in the incitement to epistemological violence that it provides. Every readerly challenge to Suzanne's credibility seems to reinforce her clamant Suzanne-ness; every impulse to impeach her, to expose her and then "show *her*," to unveil her manipulative pretensions consolidates the illusion that there is an object for these resisted and resistant epistemological desires.

And it consolidates more than that. Reality effects within the text are not the only ones enforced by the text: the reality, the solidity, the reification not only of a character but of, for instance, sexuality itself, is also enforced in, on, and then ultimately *by* the resisting reader. The doggy, fascinated lingering—Suzanne's, but hence also the novella's and ours—at the very precisely unnamed threshold of delineation between "the sexual" and "the nonsexual" does in fact a great deal to create the very threshold effects that operate so strongly there.

The latent energy in a reader's response to Suzanne at her third convent comes from the fact that Suzanne has penetrated so very far into the territory of "sexual experience" without apparently having acquired with it the accountability—to anyone, including herself—that is part of the whole harness apparatus of possessing a language of sexuality. Like the Superior, but for less explicable reasons, the reader may be anxious (not necessarily to say eager) to see her hauled back to the border and fitted out properly. The reader identifies Suzanne's experience as unequivocally "sexual" in the first place because of its rhythms of interruption and oblivion—the Superior's convulsions, the uncontrollable sleepiness of both women after their encounters—as if it were the strong and sudden demarcation of a difference between consciousness and something else (perhaps almost anything else) that made sex sex. ("Why," Suzanne demands of the Superior accusingly, "when I came back to my room after being with you, was I so upset and abstracted? . . . . Whence came a sort of lethargy I had never felt before? Why should I, who have never slept in the daytime, feel myself drifting off into slumber?" [p. 166].) Both the reader and, belatedly, Suzanne learn most, however, about the strength and importance of sexuality, not from either witnessing or experiencing any number of (female) orgasms or blackouts, but from the increasingly pointed use or anticipation of (male) interdiction: the plot by which certain touches, certain rhythms, certain opacities of consciousness come to be named as "the sexual" is precisely coextensive with the plot of the infiltration of the convent by the clerical word No.

The moment when the consolidation of "the sexual"—"the sexual" *as knowledge*—is most dramatically asserted in the story is also the moment when its entire evacuation of knowledge *content* is most startlingly evident. Only pages from the end of the text, Suzanne deliberately eavesdrops on an exchange between the Superior and her confessor:

> The first words I heard after a long silence made my blood run cold. They were:
> "Father, I am damned for ever. . . ."
> I recovered and I listened; and the veil which until then had obscured the peril I had been in was just being torn asunder when somebody called my name. I had to go and I went, but alas! I had heard all too much. What a woman, Sir, what an abominable woman!

> (Here the memoirs of Sister Suzanne become disconnected, and what follows is only notes for what apparently she meant to use in the rest of her tale. It seems that the Superior went mad, and the fragments I am about to transcribe must refer to her unhappy state.) (p. 179)

Indeed, from this point what had been a continuous narrative by a single first-person informant becomes fragmentary, incoherent, and, as it were, haunted by the disintegrative presence of that second first person ("*I* am about to transcribe"? Who?) who appears nowhere else in the book. The effects of this revelation on the Mother Superior, whose degeneration is startlingly rapid; on Suzanne, whose flight from the convent and subsequent ruin follow at once on the heels of that; and on the novella itself, which explodes into narrative shrapnel exactly here and ends very soon hereafter, are all irresistible testimonials to the power of sexual knowledge. And yet there is literally *no* information that *could* constitute this threshold of knowledge for Suzanne, that she has not already received or thought. It is entirely impossible to reconstruct what she can have overheard here. Not only has she both seen and experienced orgasmic sensations with the Mother Superior, but she has repeatedly mused on whether they might be prohibited, how prohibited they might be, reasons why they might be prohibited; in the process of actively repelling sexual "knowledge," she has done a thorough survey of the territory where that "knowledge" might live, and only her refusal ever to allow anyone to attach a name to anything differentiates her state from that of the most deeply endued initiation. Yet even the hearing of those names—say, the word "lesbian" or an eighteenth-century French surrogate—could pose no threat to the cognitive Teflon of Suzanne's ideation, since she has allowed herself to accumulate no mental deposit of connotation or denotation to which the name could attach; and without that, it might as well be a nonsense syllable. Suzanne needs—

but more importantly, the narrative impetus of the novella needs—a scene, here, of revelation, and the scene occurs. From the vantage point of afterward, what has been most achieved is the illusion of existence of the thing revealed.

In fact, the delineation of "the sexual" in this convent, in this reading, is done by a process that resembles gravestone-rubbing. The dense back-and-forth touch of the crayon leaves a positive map not of excrescences but of lines of absent or excised matter. And the pressure of insistence that makes a continuous legibility called sexual knowledge emerge from and take the shape of the furrows of prohibition or of stupor is, most powerfully, *the reader's* energy of need, fear, repudiation, projection.

The "content" of these energies, however, must not be too quickly assumed; what by the reader is needed, feared, repudiated, projected, too, is being named and given substance, though never for the first time, through the same process of reading. I see four kinds of energy content that may or must be circulating forcibly in the readerly animation of this text: sexual desire, sexual ignorance, ignorance that is not in the first place sexual, or some relation that can only uneasily be condensed either as "sexual" or as "ignorance."

The accusatory investment of Suzanne with a displaced version of an already more or less defined and reified—insofar as denied and resisted—sexual desire that emanates from the reader may perhaps be said to characterize a masculine relation to this novella. I put it this way, not because I consider all sexual desire for women to belong "properly" or definitionally to men, but because it is in this readerly relation that the distinctively patriarchal triangular structure through which *La Religieuse* enacts the male traffic in women becomes visible. Jay Caplan has written best about the currents of male ownership, tenderness, complicity, and subordination that link author and (male or male-impersonating) audience through the body and sufferings of Suzanne: "This novel has the form of a message addressed by one father to another about their symbolic (and hence, absent) daughter. . . . I am not sure whether it is more important to emphasize the incestuous character of the father-daughter relationship here or the homosexual character of the discursive situation." [13]

13. Jay Caplan, *Framed Narratives: Diderot's Genealogy of the Beholder*, afterword by Jochen Schulte-Sasse, Theory and History of Literature, vol. 19 (Minneapolis: University of Minnesota Press, 1985), p. 57. I would prefer to substitute "homosocial" for "homosexual" in Caplan's formulation. This reading of the male homosocial trajectory of *La Religieuse* is based on the novella's supposed origin in a hoax: Diderot claims that he and his group of male friends in Paris "invented" the unwilling nun, Suzanne Simonin, as a bait with which to

It is possible, on the other hand, that the projectile and circulatory energies of ignorance per se, including sexual ignorances, are less crucially gendered ones. The energies of ignorance always make an appeal to, and thus
require the expulsion of, *a time before,* a moment of developmental time,
and why not call it for convenience's sake a time before gender—factitious
though that time must be. I don't know in trying to summon up an image of
these energies whether it will be more effective to evoke in academic readers
the time before we became literate or the time before we became expert
at interpreting the signs associated with sexuality. The former is no doubt
safer and stabler for an audience of academic readers, since as few of us are
likely after all to consider ourselves even yet quite expert in the syntax and
semantics of sexual codes as we are likely to be living with an unacknowledgeable secret of illiteracy. But the two processes of socialization, each far
more protracted and painful than we are accustomed to admit—far more
so, for most of us, than for instance toilet training—do require and impel similar engines of motility: "knowing" how to read, "knowing" how
to interpret sexual meanings, both involve acrobatic leaps of yet unearned
identification consolidated by recoils of a more violent repudiation. (Perhaps I even appeal with a special confidence to the force of this energy in an
audience for whom its inexhaustible renewal is both a qualification and a
professional deformation: if the energy of our fear of, loathing for, disiden-
tification from, and at the same time fascination with the ignorance that
is only barely not our own were to evaporate, what force on earth would
drive the red pen through the papers?) I do wonder, however, whether for
heterosexual women as another specific group the costly suppression from
memory of a whole narrative of the enforced, entitling acquisition of liter-
acy may not have a particularly insistent resonance in another: the suppression, for each, to herself of the narrative of her acquisition of the "natural"
desire for men—a long (far beyond childhood) and no doubt odd process
in women or in men, but one whose compulsoriness and cultural centrality
in women seems to exact the more scouring and extortionate mystification.

Yet it would not be enough to say that it is my fear of my own sexual

----

lure back to Paris their friend, the Marquis de Croismare, who had withdrawn to his country
estate; and even more dramatically, that the suffering woman so conjured into existence as a
lure between men had then to be killed off when Croismare responded with an unpredictable
warmth. See the "Préface-Annexe" to *La Religieuse,* attributed to Friedrich Melchior Grimm,
in Assézat, *Oeuvres complètes de Diderot,* vol. 5 (*Belles-lettres* II), pp. 175–204. For an excellent discussion of theoretical issues raised by the extremely slippery textual history of this account, see Rosalina de la Carrera, "Epistolary Triangles: The *Préface-Annexe* of *La Religieuse*
Reexamined," *The Eighteenth Century: Theory and Interpretation* 29 (Winter 1988): 263–80.

desire for Suzanne that makes me propulsively individuate her as an "other" in my reading of this book. Nor is it my relation to something I can *stably* define as my own, now expelled or even only barely extruded, ignorance. That woman whose blood is clotting in my underclothes: have I indeed an effectual power to love, to legitimate, or to support her? This continues to be a question I sickeningly can't answer. The ground of its asking, even, slides from the emotional toward the economic toward the genital toward the informational . . . . Its hypostatization as either the question of a discrete sexual desire or the question of ignorance is a distancing wish, perhaps a necessary but a mutilating one. How much the more so, in this formative text of "sexualizing" literature, the conflation of those two questions in turn in the single very hard currency: *sexual ignorance*. Under this regime, it becomes exciting to be near people who may not know themselves sexually, and the excitement is very specific: it is an excitement to violence, the violence of epistemological enforcement.

There are, then, women's and men's stories, and no doubt more than these, to tell about the violence of character construction in *La Religieuse*, which is not distinct from the violence of sexual construction—that is to say, the violence by which, in the latter eighteenth century, sexuality *is* constructed. Although the novella unfolds a drama of sexual constitution distinctively for readers who identify through either gender, and although each drama pulls any reader into identifications that cross the reader's lines of gender, the bonds of desire that root the story and that create and register sexuality in it are of same-sex desire, the liminality of *its* absence or presence. This explicit definitional centering for and among all sexualities of a same-sex desire is a possibility that was repressed though far from erased in the succeeding centuries. The completed paranoia of nineteenth-century male homophobia intervened on it, followed by the coercively incoherent panoply of taxonomic discourses of male and female "homosexuality" from the turn of the century. Psychoanalysis has encrypted the definitional centrality of same-sex desire, preserving its place while carefully and repeatedly stepping over the place unnamed; and it has remained for twentieth-century feminist and gay movements to place it, agonistically, under conscious recovery, but in a landscape already deeply marked by its minoritization.

It is a chief motive for the study of the epistemology of the closet, as indeed it is part of its implicit axiomatic grounding, that a defining feature of twentieth-century fascisms—fascisms past and fascisms perhaps to recur—will prove to have been a double ideological thrust along the axis of same-sex bonding or desire. The first such thrust, characteristic in any case of the twentieth century and in our decades geometrically increasing, is a very heightened foregrounding of same-sex bonding (especially male) to the individual and societal mind: heightened in its visibility, in its perceived problematicalness, and not least importantly in its investment with a charge specifically of "sexuality" and of sexual representativeness and sexual knowledge.[14]

14. The dangerously homophobic folk wisdom now endemic in both high- and middlebrow culture that sees in the rouged cheeks of Gustav von Aschenbach, or of Joel Grey in *Cabaret*, a sexualized "decadence" from whose image the supposedly answering image of fascism is seen as inseparable, finds too ready an echo in the popular pornographic fantasy, in the photogenic anti-Nazi film fantasy (of *The Damned, The Conformist, The Fighters*), and in the currently respectable homophobic feminist-theory fantasy, that join in hallucinating that (as Maria-Antonietta Macciocchi asserts) "the Nazi community" itself "is made of homosexual brothers"—or, in Richard Plant's trenchant paraphrase, "that the incomprehensible Nazi crimes could be easily explained: the Nazis were simply homosexual perverts." (Plant's discussion of the three films cited [*The Fighters* was a Soviet film made in 1936 by Gustav von Wangenheim], including the sentence I quote, are in his indispensable book, *The Pink Triangle: The Nazi War Against Homosexuals* [New York: Henry Holt, 1986], pp. 15–16.) Macciocchi's formulation is quoted in Jane Caplan, "Introduction to Female Sexuality in Fascist Ideology," *Feminist Review* 1 (1979): 62. It is congruent, more broadly, with the identification (in the work of, for example, Irigaray) as flatly, transhistorically "homosexual" of the male homosocial bonds that largely structure patriarchal culture. Precisely to the degree that this is a potent polemical move, it is a dangerous and demagogic one: its rhetorical kick depending on and hence reinforcing our own historically specific culture's distaste, not in the first place for patriarchal oppression, but for homosexuality itself. (Craig Owens gives a fuller account of, and argument against, this strain in feminist theory in "Outlaws: Gay Men in Feminism," *Beyond Recognition: Representation, Power, and Culture* [Berkeley: University of California Press, 1992], pp. 218–35.) There are, at any rate, far more scrupulous, admirably discriminating, though less theoretically sweeping feminist investigations of fascism under way, such as those represented in Renate Bridenthal, Atina Grossmann, and Marion Kaplan, eds., *When Biology Became Destiny: Women in Weimar and Nazi Germany* [New York: Monthly Review Press, 1984].) It should be unnecessary to say that the fantasy of Nazi homosexuality is flatly false; according to any definition of homosexuality current in our culture, only one Nazi leader, Ernst Roehm, was homosexual, and he was murdered by the SS on Hitler's direct orders in 1934. What seems more precisely to be true is that at any rate German fascism (like, in less exacerbated form, twentieth-century Western culture at large) emerged on a social ground in which "the homosexual question" had been made highly salient. As much gay historiography now shows, the new salience of "the homosexual question" can be tied to a variety of forces including the complex changes in medical and psychiatric discourse, new and incoherent mappings of homo/heterosexual definition both in relation to minority identity and in relation to masculinity/femininity, conflicting forms of an unprecedented homosexual activism, developments of xenophobic nationalism, and preoccupying homosexual scandals in England and Germany. Implicit in that ambiguous but conveniently encapsulated notion of "the homosexual question," as in, for instance, that of "the Jewish

The second ideological thrust I have mentioned seems to go against the first, but, far from neutralizing or undoing it, is in fact the absolute precondition, the necessary activating structure, for the recuperation of its passional energies to the sustained uses of any fascism. That is the maintenance of an almost unbreached separation between the heightened surcharge of the homosocial/homosexual on the one hand, and on the other hand any availability across the society of values or language or worldviews that would explicitly allow these strong charges to be respected, felt through, legitimated, and inhabited, not to say loved. Fascism is distinctive in this century not for the intensity of its homoerotic charge, but rather for the virulence of the homophobic prohibition by which that charge, once crystallized as an object of knowledge, is then denied *to* knowledge and hence most manipulably mobilized.[15] In a knowledge regime that pushes toward

---

question," are both the potential for expression of diametrical oppositions of opinion, and at the same time an overarching tendentiousness of address (why should the question considered not be "the heterosexist," "the anti-Semitic," or more simply "the heterosexual," "the Christian" question?) so grave that to attain the status of a "question" necessarily both records and aggravates the endangeredness of any group. The political bearings of "the homosexual question" in Germany were for a long time unsettled and contradictory, especially insofar as they intersected the fascist intensification of gynephobia and male homosocial bonding. Indeed, while most of the early gay rights discourse was strongly antifascist, it appeared for some time that the homosocial heightening involved in early fascism might offer potent affordances to at least some forms of homosexual advocacy. Roehm's declared political and military heroes, for instance, were historical figures, many homosexual, of whom he said "One can barely imagine that they yielded to feminine wiles"; and Hitler for a while supported Roehm by demarcating and protecting from scrutiny the "private domain" of homosexual SA officers lest his militia turn into "an institute for the moral education of genteel young ladies." The formative male-male intensities of the politically important pre-Nazi, in some respects proto-Hitler Youth *Wandervogel* movement, similarly, were available for ideological investment by the masculist, militarist homosexual advocate Hans Blüher. (Plant, *Pink Triangle*, pp. 60, 61 on Roehm and the SA; pp. 42–43 on Blüher.)

15. For instance, in the Nazi case: Roehm was murdered early on. Blüher, despite his embrace of the right, was reviled by the Nazis when they came to power, and as Richard Plant explains, "the Youth Movement was first absorbed and then destroyed by Hitler Youth leaders, when its 'decadent' and 'elitist' homoeroticism succumbed to the [highly congruent] 'racially productive' blood-and-soil philosophy of the Nazis." The differences between the two positions had to do with the fascist hypostatization of the *name of the family*, not with the actual practicalities of breeding: in Blüher's reading of history, as in the fascist, a reproductive heterosexuality had been compatible for soldierly men with a hyper-charging of male homosocial bonds (Plant, *Pink Triangle*, p. 42). Nazis gave the highest priority to the destruction of Magnus Hirschfeld's and other antihomophobic archives, research, and writing, and to broadening widely and indefinitely the legal boundaries of "criminal indecency between men." Finally, in Plant's cautious estimate, between 5,000 and 15,000 men were murdered in concentration camps for being homosexual (*Pink Triangle*, pp. 152–81).

the homosexual heightening of homosocial bonds, it is the twinning with that push of an equally powerful homophobia, and most of all the enforcement of cognitive impermeability between the two, that will represent the access of fascism.

I don't know if there was ever a time when the impulse against the sexual definition and charging of same-sex bonds could have been politically progressive. (Though it seems likelier, and the possibility more recent, in the case of women's bonds than of men's.) If so, at any rate, that time is past. Under the accomplished and consolidated regime of "sexual knowledge," a move toward sexual de-charging can become manifest as nothing but the repressive appeal to a modern origin-myth of primeval sexual innocence. Such a move is by now fatally appropriable as a move toward sexual delegitimation, toward the cognitive and ideological apartheid around homosexuality that will provide the undergirding of any new fascism. It is only with this understanding that the political concept of a fight against sexual ignorance can make sense: a fight not against originary ignorance, nor for originary innocence, but against the killing pretense that a culture does not know what it knows. The only move that I can see worth making in this context is the actively antihomophobic one, valuing and exploring and sharing a plurality of sexual habitation, love, and even crucially knowledge. Yet it can be done only with every possible sophistication about the exclusionary and inflictive involvements of that knowledge.

"Privilege of Unknowing" was written in 1986. I was helped in my work on and around the essay by discussions with and responses from Rosalina de la Carrera, Jonathan Dollimore, Michael Fried, Joan Liberman, and D. A. Miller.

TALES OF THE AVUNCULATE:

QUEER TUTELAGE IN *THE*

IMPORTANCE OF BEING EARNEST

> *Through his memories I recall hours on end*
> *sitting in the weeds in the backyard next to*
> *the lawn chair where my uncle lay in shorts*
> *and a wedding ring, his body hardened and*
> *brown from days of skin diving in faraway*
> *oceans filled with the mysterious fish and crea-*
> *tures he described. I stared and stared and some-*
> *times played with his arms for hours and I remem-*
> *ber feeling a slight dizziness that years later I came*
> *to see first as a curse and then as a tool: a wedge that*
> *I might successfully drive between me and a world that*
> *was rapidly becoming more and more insane.*—David
> Wojnarowicz[1]

> *There have always been aunts in my family, and Uncle*
> *Willie also had his aunts.*—E. M. Forster[2]

Let's begin—but only because everyone else does—with
the Name of the Father. *The Importance of Being Earnest* is
famous for ending with a scene in which its hero, who never
knew his father (having been found as a baby in a handbag in
the left luggage office of Victoria Station), tries to ascertain his
parentage from his newfound aunt, Lady Bracknell, in order to satisfy a girlfriend who has determined that she can only marry a man
named Ernest.

---

1. David Wojnarowicz, *Close to the Knives: A Memoir of Disintegration* (New York:
Vintage Books, 1991), p. 80.
2. E. M. Forster, "Uncle Willie" (a manuscript essay about the uncle on whom Forster had
based the figure of Rickie's aunt in *The Longest Journey*); reprinted in *The Longest Journey*, ed. Elizabeth Heine (London: Penguin Books, 1984), pp. 294–300; quoted from p. 294.

JACK: Lady Bracknell, I hate to seem inquisitive, but would you kindly inform me who I am?

LADY BRACKNELL: I am afraid that the news I have to give you will not altogether please you. You are the son of my poor sister, Mrs. Moncrieff, and consequently Algernon's elder brother.

JACK: . . . Aunt Augusta, a moment. At the time when Miss Prism left me in the hand-bag, had I been christened already?

LADY BRACKNELL: Every luxury that money could buy, including christening, had been lavished on you by your fond and doting parents.

JACK: Then I was christened! That is settled. Now, what name was I given? Let me know the worst. . . .

LADY BRACKNELL (*meditatively*): I cannot at the present moment recall what the General's Christian name was. But I have no doubt he had one. He was eccentric, I admit. But only in later years. . . .

JACK: His name would appear in the Army Lists of the period, I suppose, Aunt Augusta?

LADY BRACKNELL: The general was essentially a man of peace, except in his domestic life. But I have no doubt his name would appear in any military directory.

JACK: The Army Lists of the last forty years are here. These delightful records should have been my constant study. (*Rushes to bookcase and tears the books out.*) M. Generals . . . Mallham, Maxbohm, Magley, what ghastly names they have—Markby, Migsby, Mobbs, Moncrieff! Lieutenant 1840, Captain, Lieutenant-Colonel, Colonel, General 1869, Christian names, Ernest John. (*Puts book very quietly down and speaks quite calmly.*) I always told you, Gwendolen, my name was Ernest, didn't I? Well, it is Ernest after all, I mean it naturally is Ernest.[3]

No reader of recent critical theory will find it hard to imagine the joy of recognition that such a passage induces—as if in imitation of the play's own denouement—in sophisticated critics, like Christopher Craft who recognizes in Lady Bracknell "a deconstructionist before her time, a proper Derrida in late Victorian drag."[4] In a condensation with which Derrida or Lacan would feel equally at home, the Name of the Father seems here, all but explicitly, exposed as the guarantor and enforcer of two things simultaneously: the fiction of a "natural" correspondence between names and

---

3. Oscar Wilde, *The Importance of Being Earnest and Other Plays*, introduction by Sylvan Barnet (New York: New American Library, 1985), pp. 187–89.

4. Christopher Craft, "Alias Bunbury: Desire and Termination in *The Importance of Being Earnest*," *Representations* 31 (Summer 1990): 21; further cited in the text as CC.

things, signifiers and signifieds, titles and selves; and by the same token the forcible imposition of a "natural" narrative telos in heterosexual marriage and the family. As Craft puts it, Wilde

> expressly targets the most overdetermined of . . . signifiers—the Name of the Father, here Ernest John Moncrieff—upon whose lips . . . a whole cultural disposition is hung: the distribution of women and (as) property, the heterosexist configuration of eros, the genealogy of the "legitimate" male subject, and so on. Closing with a farcical pun on the father's name, Wilde discloses, in a single double stroke, the ironic cathexis (and the sometimes murderous double binding) by which the homosexual possibility is formally terminated . . . . So decisive is the descent of the father's name, so swift its powers of compulsion and organization, that (at least seemingly) it subdues the oscillations of identity, straightens the byways of desire, and completes—*voilà*—the marital teleology of the comic text. (CC, p. 36)

In Craft's reading, as in Joel Fineman's influential one of 1980,[5] any outlets from this monolithic cultural imposition lie in the strains that can be painstakingly traced within the pseudonatural propriety of the name—to begin with, of course, the proper and paternal name. For Craft, as also for Jonathan Dollimore, the occluded gay possibilities of the play, what are literally (in the homophobic speech prohibition dating back to St. Paul) *non nominandum,* not to be named, surface mainly as an oscillation in the relation *to* the name: and in the relation, specifically, of self to name. Finally, in the very concept of self. Craft: "For what Wilde seeks in desire is not the earnest disclosure of a single and singular identity . . . but rather something less and something more: the vertigo of substitution and repetition" (p. 22).

The vertiginous oscillation of "same" and "different" is the sensation most stably valued by this reading, and the one most identifiable with homosexual being. On the phonemic level, Craft argues, punning itself "becomes homoerotic because homophonic. Aurally enacting a drive toward the same, the pun's sound cunningly erases, or momentarily suspends, the semantic differences by which the hetero is both made to appear and made to appear natural, lucid, self-evident" (p. 38). Dollimore, correspondingly, sees in Wilde's propensity for *grammatical inversion* a destabilization of

5. Joel Fineman, "The Significance of Literature: *The Importance of Being Earnest,*" *October* 15: 79–90.

the essentialist category "self" that has an intimate relation, as well, to what he describes as Wilde's experience of *sexual* inversion.[6]

Each of these readings traces and affirms the gay possibility in Wilde's writing by identifying it—feature by feature, as if from a Most Wanted poster—with the perfect fulfillment of a modernist or postmodern project of meaning-destabilization and identity-destabilization. There is no question that the play, like Wilde's other writing, answers spectacularly to any such interrogation. It seems as if Wilde and his sexuality are being strongly validated, in these readings, through the almost uncanny verbal mirrorings they offer to certain already prestigious theoretical projects "after realism," "after Freud," "after metaphysics"—projects that have not themselves, however, had gay affirmation or an antihomophobic problematic at their heart; arguably quite the contrary. Wilde's sexuality seems to underwrite discoveries of which we have already heard much in the register of heterosexism: a utopian aesthetic of the dizzying, an Oedipally centered demonstration of Oedipal impossibility. We are to admire Wilde for being Derrida or Lacan *avant la lettre;* "inversion" and the "homosexual" are hailed as magically exact precursor-supplements to a line of modernist/ postmodern phantasmatic. Magically exact but perhaps not coincidentally exact—if, as one might want to argue, not only have the identities of the "invert" and the "homosexual" already been intimately marked, historically, by the requisitions of the very same modern phantasmatic; but, as well, that phantasmatic has itself already been intimately marked by the expulsion of the homosexual destined so neatly to arrive as its supplement.

Thus, on the one hand I find these deconstructive readings of Wilde indispensably interesting and, to an almost tautological degree, "true." There is no question that Wilde is up to what these critics see him as being up to; how could he help being so, if both he and the violently repressive energies that silenced him and filled up the vacuum of his absence were so written into the origin of such interrogations? On the other hand, it also seems as urgent as it is difficult to find some alternative approaches: angles from which it might be possible to perceive the less theorizable resistances that Wilde may at the same time have been offering to these homogenizing modern(ist) interpretive projects.

I would suggest, to give only one reason for my wariness about these deconstructive celebrations, that Wilde "as a person" does not make it par-

6. Jonathan Dollimore, *Sexual Dissidence: Augustine to Wilde, Freud to Foucault* (New York: Oxford University Press, 1991), pp. 14–15.

ticularly easy to assimilate his own sexuality, even insofar as it is oriented toward other men, to either of the then newly available models for male-male sexuality, "inversion" or "homosexuality," the models that so enticingly facilitate the deconstructive project. There is no evidence of Wilde's vibrating very strongly to the chord of gender inversion—to the trope of the woman's soul trapped in a man's body, in the famous 1869 phrase of Karl Heinrich Ulrichs. When he did appeal to its energies, it was always with some extra twist that would make nonsense again out of the supposed schematizations of this model—as when it's *Gwendolen* who pronounces, "The home seems to me to be the proper sphere for the man. And certainly once a man begins to neglect his domestic duties he becomes painfully effeminate, does he not? And I don't like that. It makes men so very attractive" (p. 162). Although Wilde's ambience was in many exciting ways charged with forms of gender transitivity, he does not seem very much to have seen or described either himself or those he loved in terms of inversion; although the critics' assimilation of his characteristic grammatical inversions with a play of gender inversion would seem to depend on some such self-conception. What they depend much more on, however, is the constitution of gender inversion as a transparent and unexaminable part of the "common sense" of twentieth-century sexual tropology—however uneagerly the eros of Wilde himself may answer to such interpretation.

But neither, however, do the associations of the late-nineteenth-century coinage "homosexuality" appear to have engaged much of Wilde's formidable propriodescriptive energy. As David Halperin describes some of the concomitants of "homosexuality" (as distinguished from "inversion"):

> The conceptual isolation of sexuality *per se* from questions of masculinity and femininity made possible a new taxonomy of sexual behaviors and psychologies based entirely on the anatomical sex of the persons engaged in a sexual act (same sex vs. different sex); it thereby obliterated a number of distinctions that had traditionally operated within earlier discourses pertaining to same-sex sexual contacts . . .: all such behaviors were now to be classed alike and placed under the same heading. Sexual identity was thus polarized around a central opposition rigidly defined by the binary play of sameness and difference in the sexes of the sexual partners; people belonged henceforward to one or the other of two exclusive categories. . . . Founded on positive, ascertainable, and objective behavioral phenomena—on the facts of who had sex with whom—the new sexual taxonomy could lay claim to a descriptive, trans-historical validity. And so it crossed

the "threshold of scientificity" and was enshrined as a working concept in the social and physical sciences.[7]

Initiating, along with the stigma of narcissism, the utopic modern vision of a bond whose egalitarian potential might be guaranteed by the exclusion of any consequential difference, the new calculus of homo/hetero, on which critics draw in assimilating Wilde's "homo" sexuality to his use of the "homophonic" pun as opposed to "heterophonic" reference (as Craft does, p. 38), or of the "autological" as opposed to the "heterological" signifier (as Fineman does, pp. 88–89)—the new calculus, as I have pointed out in *Epistemology of the Closet,* owes its sleekly utilitarian feel to the linguistically unappealable classification of anyone who shares one's gender as being "the same" as oneself, and anyone who does not share one's gender as being one's Other.

It is startling to realize that the aspect of "homosexuality" that now seems in many ways most immutably to fix it—its dependence on a defining *sameness* between partners—is of so recent crystallization. (The process is also, one ought to add, still radically incomplete and geoculturally partial.) Wilde's work was certainly marked by a grappling with the implications of the new homo/hetero terms. I argue in *Epistemology of the Closet,* for instance, that this is one way of describing the trajectory of *The Picture of Dorian Gray:* the novel takes a plot that is distinctively one of male-male desire—the competition between Basil Hallward and Lord Henry Wotton for Dorian Gray's love—and condenses it into the plot of the mysterious bond of figural likeness and figural expiation between Dorian Gray and his own portrait. The suppression of the original defining *differences between* Dorian and his male admirers—differences of age and initiatedness, in the first place—in favor of the problematic of Dorian's *similarity to* the painted male image that is and isn't himself, seems to reenact the discursive eclipse in this period, by the "homo"-sexual model, of the Classically based *pederastic* assumption that male-male bonds of any duration must be structured around some diacritical difference—old/young, for example, initiator/initiate, or insertive/receptive—whose binarizing cultural power would be at least comparable to that of gender.

But it seems patent that in the paradigm clash among these ways of thinking male-male desire, Wilde's own eros was most closely tuned to the note of the pederastic love in process of being superseded—and, we may as well

7. David M. Halperin, *One Hundred Years of Homosexuality* (New York: Routledge, 1989), p. 16.

therefore say, radically misrepresented—by the homo/hetero imposition. Though a passionate classicist, Wilde did not desire only boys; but his desires seem to have been structured intensely by the crossing of definitional lines—of age, milieu, initiatedness, and physique, most notably—sufficiently marked to make him an embattled subject for the "homosexual" homo-genization that is by now critically routine. It is routine, to repeat, not because of its adequacy to Wilde's desiring self—though it was certainly part, a resisted part, of his discursive and creative world—but rather routine because it so efficiently fills so many modern analytic, diagnostic, and (hence) even deconstructive needs.

As we have seen, the indispensable—but, I am arguing, insufficient—deconstructive reading of *Earnest* always seems, like the play's hero, to have its origin in a terminus. It doesn't pass Go; it doesn't collect $200; it heads straight for the end-of-the-third-act anagnorisis (recognition or de-forgetting) of the Name of the Father. The sexual different-ness of the play and/or of its author gets subsumed in these readings, under the law of the Father, as that one-size-fits-all "difference" that can always be conscripted to play the same old play with the "same." Bottom line: the totemic force of the Oedipal father-son imperative, its systemic equivalence with everything that could at all be called family, individuation, or meaning, is only strengthened by the congruence with it of these new and glamorous, notionally subversive, terms and sexualities—by the rhetorical sublimity and "dizzying" unthinkability of any alternative topos.

Forget the Name of the Father.

Forget the Name of the Father!

Why can't that, which is after all what these characters remarkably *do* for the first seventy-four seventy-fifths of *The Importance of Being Earnest,* be said to constitute its imperative—that, rather than the final forced march of the play's amnesiac farce into the glare-lighted, barbed-wire Oedipal holding pen of the very last page?

The injunction to forget, of course, to forget something-in-particular after its jolting anamnesis, as you know if you've ever tried to do it—four in the morning, haplessly alert, knowing you'll never get back to sleep if you can't stop thinking about a certain X—opens an interminably self-defeating involution; self-defeating unless some other term (if only some other fetishistically summoned image, another name, another face, another repeated phrase) can be substituted for X. The only way to forget X is to invoke Y in her place. It can only partially work (your obsessional thought will not cease to be marked by the structuring imperative of X; by analogy, I hardly suppose a coherent textual approach to be possible that would lie fully outside the terms of the anagnoritic ones) but it can work "enough," in the sense of making an unforeseen difference in the effect to which X presides over your obsessional process. I suggest, or I suggest the play suggests: Forget the Name of the Father. Think about your uncles and your aunts.

Think about aunts and uncles because, to begin with, the presiding representative of the previous generation in *The Importance of Being Earnest* is neither the heroes' father nor even their mother but an aunt, Lady Bracknell—a part often played, of course, *Charley's Aunt*-style, by a male actor in woman's dress. Although Aunt Augusta is the very opposite of effeminate, "aunt," "auntie," or the French "tante" were recognized throughout the nineteenth century, and are still widely recognized, as terms for (what an 1889 slang dictionary calls) a "passive sodomist"—or, more likely, for any man who displays a queenly demeanor, whatever he may do with other men in bed.[8] (Proust's original name for his 1909 essay on "the Men-Women of Sodom," the real catalyst, apparently, of *Remembrance of Things Past*, was more simply: "La Race des tantes.") "Uncle," at the same time, was a common term for a male protector in a sexual relation involving economic sponsorship and, typically, class and age transitivity. "Uncle" has been common, as well, in gradations from the literal, as a metonym for the whole range of older men who might form a relation to a younger man (as patron, friend, literal uncle, godfather, adoptive father, sugar daddy) offering a degree of initiation into gay cultures and identities—like the older man whom a friend of mine, my age, always refers to warmly as his fairy godmother.

"Uncle" and "aunt" in these very gay-marked meanings don't add up to two complementary male roles, as for instance a "masculine" and a "feminine": even if you wanted to, you couldn't pair an uncle up with an auntie and bundle them off for a happy, heterosexually intelligible honeymoon. Uncle and aunt aren't even both in the same sense "figures of speech." Furthermore "aunt," used about a man, alludes to a gender-transitive persona which, however, it doesn't particularly pretend to stabilize in the dyadic terms of gender inversion: the "aunt" usage long predates and surely influences, but is not adequated by, the rationalized discursive production of the invert.[9] "Aunt" tells who or how you are (at least sometimes) but not

8. Quoted in Neil Bartlett, *Who Was That Man?: A Present for Mr. Oscar Wilde* (London: Serpent's Tail, 1988), p. 90, from Barbère, *Argot and Slang Dictionary* (1889).

9. Proust's section on "La Race des tantes," claiming to explicate the hidden calculus of inversion under the rubric of the Aunties, is famous as a thicket of pseudo-scientific self-contradiction. (Marcel Proust, "La Race des tantes," one of the provisional titles for "La Race maudite," in *"Contre Sainte-Beuve" suivi de "Nouveaux mélanges,"* 14th ed. [Paris, Gallimard, 1954].) See, on this, my *Epistemology of the Closet*, pp. 217–23.

whom you desire. "Uncle" is very different, *not* a persona or type but a relation, relying on a pederastic/pedagogical model of male filiation to which
also, as we have seen, the modern rationalized inversion and "homo-"
models answer only incompletely and very distortingly. But of course it is
the very badness of the fit of uncle and aunt—the badness of their fit with
each other in the first place, but also with the streamlined modern models of
"family" and of same-sex attachment—that makes them such good places
to look for some of the gravity of Wilde's resistance to the sleek "same"/
"different" scientism of modern gender and sexual preference.

The interest of uncles and aunts isn't confined to particular, more or less
figural usages of the words, though. There are plenty of signals that the
constitution and recognition of aunts and uncles has as much as that of
parents to do with the identity issues of *The Importance of Being Earnest*.
Early in the play, for example, there's a scene where Jack tries to wheedle
his friend Algernon into returning his missing cigarette case, without acknowledging to Algernon the existence of his marriageable young ward
Cecily, from whom it was a gift.

> JACK: I simply want my cigarette case back.
> ALGERNON: Yes; but this isn't your cigarette case. This cigarette case
> is a present from someone of the name of Cecily, and you said you
> didn't know anyone of that name.
> JACK: Well, if you want to know, Cecily happens to be my aunt.
> ALGERNON: Your aunt!
> JACK: Yes. Charming old lady she is, too. Lives at Tunbridge Wells.
> Just give it back to me, Algy.
> ALGERNON: But why does she call herself little Cecily if she is your
> aunt and lives at Tunbridge Wells? (*Reading.*) "From little Cecily with
> her fondest love."
> JACK: My dear fellow, what on earth is there in that? Some aunts are
> tall, some aunts are not tall. That is a matter that surely an aunt may
> be allowed to decide for herself. You seem to think that every aunt
> should be exactly like your aunt! That is absurd! [*Four-act version
> adds here:* There is a great variety in aunts. You can have aunts of any
> shape or size you like. My aunt is a small aunt.] [10] For Heaven's sake
> give me back my cigarette case.

10. Oscar Wilde, *The Importance of Being Earnest. A Trivial Comedy for Serious People. In
Four Acts as Originally Written* (New York: New York Public Library, 1956), 2 vols. (vol. 1
is the printed transcript, vol. 2 a facsimile), vol. 1, p. 12. Further quotations from this edition
will be noted in the text as Four-act with page number.

ALGERNON: Yes. But why does your aunt call you her uncle? "From little Cecily, with fondest love to her dear Uncle Jack." There is no objection, I admit, to an aunt being a small aunt, but why an aunt, no matter what her size may be, should call her own nephew her uncle, I can't quite make out. (pp. 120–21)

What may be at stake in the making visible of aunts and uncles in this play? In some cryptic but very provocative paragraphs in his essay "Outlaws: Gay Men in Feminism," Craig Owens suggests that the turn-of-the-century Freudian recasting of the (supposedly universal) incest taboo—from being, as anthropologists describe it, a prohibition that chiefly involves avuncular and sibling-in-law relations, to being, in the Oedipal, a prohibition of directly cross-generational relations between parent and child—may have had gravely obfuscatory consequences for modern understandings of sexuality. The possibility of an uncle-centered rather than a parent-centered reading of traditional cultures suggests, Owens says, "that the incest taboo may actually work to integrate homosexual impulses into the sexual economy, and that the 'repression' of homosexuality may be less [than] universal." [11]

In saying this, Owens may be drawing on some formulations of Juliet Mitchell's that are not themselves particularly concerned with homo/hetero issues. In *Psychoanalysis and Feminism,* Mitchell, too, points to the distance between on the one hand the broadly filiated "anthropological" incest taboo, involving men's circulation of sisters, to be rewarded by the acquisition of brothers-in-law and nephews (as formulated by, for example, Lévi-Strauss); and on the other the more nuclearized incest taboo formulated as the Oedipus complex, "its internalized form," in Freud. The difference, essentially, seems to be the dropping-out of a fourth term that had complicated and rendered "cultural" the biological triad: that fourth term being the maternal uncle. In the avuncular relation, Mitchell summarizes,

> The maternal brother offers his sister to his thereby future brother-in-law, within this generation he therefore acts as mediator between his brother-in-law and the latter's wife (his sister); furthermore, he mediates between these parents and their child (his nephew); he thus has a horizontal and vertical role. The uncle insures that the vicious cycle [of incestuous brother-sister endogamy] cannot come again. . . . The holy (bestial) family will not reign supreme. There always has to be some other term that mediates and transforms the deathly symmetry,

11. Owens, "Outlaws: Gay Men in Feminism," *Beyond Recognition,* p. 228.

the impossibility for culture, of the biological family: within the kinship structure, classically this term is the maternal uncle.[12]

Mitchell gives a Marxist-inflected account of the transition from kinship societies to modern ones:

> In economically advanced societies, though the kinship exchange system still operates in a residual way, other forms of economic exchange—i.e. commodity exchange—dominate[,] and class, not kinship structures prevail. It would seem that it is against a background of the *remoteness* of a kinship system that the ideology of the biological family comes into its own. In other words, that the relationship between two parents and their children assumes a dominant role when the complexity of a class society forces the kinship system to recede. Under capitalism the vast majority of the population not only has nothing to sell but its labour-power, it also has nothing to exchange but this. . . . Broadly speaking, kinship relationships are preserved as important among the aristocracy (a hangover from feudalism) and the cult of the biological family develops within the middle class. . . . With compulsory education, prohibition of child labour and restriction of female labour, with the increased national wealth from imperialism, the working class was gradually able to follow the middle-class example of cultivating the biological family as, paradoxically, indeed almost impossibly, the main social unit. The biological family thus becomes a major cultural event under capitalism and asserts itself in the absence of prominent kinship structures.[13]

Craig Owens's speculations invite us to ask, as Mitchell doesn't quite do, not only about the excisions involved in this avunculosuppressive move from "kinship" to "family," but also—within the postkinship culture of the now weirdly resurrected "holy (bestial) family"—about the importance of residual, re-created, or even entirely newly imagined forms of the avunculate. To begin with, of course, the fact that we don't in English so much as have names to distinguish our mother's from our father's brother (or sister), or any of those from an aunt or uncle related to us only by marriage to a parental sibling, shows that a far less specified set of avuncular roles and relations now obtains—to the extent that, first, the term "avunculate" actually does seem usable for both men and women who occupy these rela-

12. Juliet Mitchell, *Psychoanalysis and Feminism* (New York: Vintage Books, 1975), p. 375. Further page citations in the text.
13. Ibid., pp. 378–80.

tions to us; and, second, many geocultural settings allow us to call "aunt" or "uncle" people older than ourselves who aren't related to us by either blood *or* marriage.

Because aunts and uncles (in either narrow or extended meanings) are adults whose intimate access to children needn't depend on their own pairing or procreation, it's very common, of course, for some of them to have the office of representing nonconforming or nonreproductive sexualities to children. We are many, the queer women and men whose first sense of the possibility of alternative life trajectories came to us from our uncles and aunts—even when the stories we were allowed to hear about their lives were *almost* unrecognizably mangled, often in demeaning ways, by the heterosexist hygiene of childrearing. But the space for nonconformity carved out by the avunculate goes beyond the important provision of role models for proto-gay kids. After all, many of us don't turn out all that much like "artistic" Uncle Harvey, "not the marrying kind" Cousin David whose engagements always fell through at the last minute, or (the best loved people in my family) Aunt Estelle and Aunt Frances, sisters who slept in the same room for most of their eight decades. But if having grandparents means perceiving your parents as somebody's children, then having aunts and uncles, even the most conventional of aunts and uncles, means perceiving your parents as somebody's sibs—not, that is, as alternately abject and omnipotent links in a chain of compulsion and replication that leads inevitably to *you;* but rather as elements in a varied, contingent, recalcitrant but re-forming seriality, as people who demonstrably could have turned out very differently—indeed as people who, in the differing, refractive relations among their own generation, can be seen already to have done so.

It follows that a family system understood to include an avuncular function might also have a less hypostatized view of what and therefore how a child can desire. When "the family" is stylized as the supposed biologically based triad,[14] as it now is both in psychoanalysis and in the modern mass ideology that psychoanalysis in this respect both reflects and ratifies, the paths of desire/identification for a given child are essentially reduced to two: identification ("Oedipal"), through the same-sex parent, with a desire for the other-sex parent; or identification ("negative Oedipal"), through the other-sex parent, with a desire for the same-sex parent.[15] If the so far

14. I say "supposed" because there is nothing *more biological* about a family of three than a kinship group of two hundred.

15. This account is extremely oversimplified, but probably not to an invalidating degree—the recent essays of, for instance, Kaja Silverman would serve as a confirmatory limit case. Kaja Silverman, *Male Subjectivity at the Margins* (New York and London: Routledge, 1992).

undiminished reliance of psychoanalytic thought on the inversion topos were not enough to insure its heterosexist bias,[16] its heterosexist circumscription would nonetheless be guaranteed, if it is not already caused, by the fact that the closed system of "the family," within which all formative identification and desire are seen to take place, is limited by tendentious prior definition to parents—to adults already defined as procreative within a heterosexual bond.

Within this ideological system, accounts of the desire of any child must in turn be disfiguringly ritualized. The dispiriting debates on "the seduction theory," for example, have pitted a psychoanalytic-identified view of the totally volitional, unproblematically "active" child against a feminist-identified view of the child as the perfect victim, totally passive and incapable of relevant or effectual desire. But suppose we assume for a moment the near-inevitability of any child's being "seduced" in the sense of being inducted into, and more or less implanted with, one or more adult sexualities whose congruence with the child's felt desires will necessarily leave at least many painful gaps.[17] (I mean here to designate a continuum that extends to, but is not fully defined by, the experience of a child who is in fact assaulted or raped.) That a child, objectively very disempowered, might yet be seen as being sometimes in a position to influence—obviously to radically varying degrees—*by whom s/he may be seduced;* as having some possible degree of choice, that is to say, about *whose* desire, what conscious and unconscious needs, what ruptures of self and what flawed resources of remediation, are henceforth to become part of her or his internalized sexual law: such a possibility is thinkable only in proportion as the child is seen as having intimate access to some range of adults, and hence of adult sexualities. At the Utopian limit: "there is a great variety in aunts. You can have aunts of any shape or size *you like.*"

Part of the interest of the avunculate is, as we have seen, that its thinkability also renders more thinkable (across and perhaps therefore within generations) the sibling relation. *The Importance of Being Earnest* places the sibling plot in many ways prior to, and hence more tellingly in question than, the marriage plot—especially as a locus for explorations of gender

16. The inversion topos depending, after all, on a view of desire itself as something that can subsist only between a "masculine" self and a "feminine" self, in whatever sex of bodies these notional selves may be housed.

17. This account depends on both Sandor Ferenczi, "Confusion of Tongues between Adults and the Child," *Final Contributions to the Problems and Methods of Psycho-analysis,* ed. Michael Balint (New York: Basic Books, 1955) pp. 156–67; and Jean Laplanche, *New Foundations for Psychoanalysis,* trans. David Macey (Oxford: Basil Blackwell, 1989), pp. 89–149.

and sexuality. The first effect of the anagnoritic ending is not, after all, to enable the play to bring about its three heterosexual marriages, but rather to locate Jack/Ernest within a sib network.

> LADY BRACKNELL: You are the son of my poor sister, Mrs. Moncrieff, and consequently Algernon's elder brother.
> JACK: Algy's elder brother! Then I have a brother after all. I knew I had a brother! I always said I had a brother! Cecily,—how could you have ever doubted that I had a brother? (*Seizes hold of Algernon.*) Dr. Chasuble, my unfortunate brother. Miss Prism, my unfortunate brother. Gwendolen, my unfortunate brother. Algy, you young scoundrel, you will have to treat me with more respect in the future. You have never behaved to me like a brother in all your life.
> ALGERNON: Well, not till to-day, old boy, I admit [when Algernon had tried to cheat Jack]. I did my best, however, though I was out of practice. (pp. 187–88)

Only Gwendolen's prodding reminds Jack that the questions of his name and hence his marriage are also at stake. ("Good heavens! . . . I had quite forgotten that point.")

An important switchpoint between the avuncular and the sibling relations in *Earnest* is, rather oddly, the word and notion "German." But maybe not *absolutely* oddly, if we suppose that an important switchpoint between any avuncular and sibling relations is likely to be the cousin— the first cousin, more precisely called cousin german, meaning the child of a parental sibling (someone of the same "germ" as the parent), from the same germ as for instance Spanish *hermano*, brother. Of course it would, no doubt, be anachronistic to connect the Germanities of *Earnest* fully to the usage in the next decade, in the wake of the Eulenburg scandal in Germany, of "Do you speak German?" as a pickup line for gay Frenchmen; in 1895 homosexuality is not yet referred to as *le vice allemand*. But German is already, as Miss Prism enthuses in the four-act original of *Earnest,* famous as the tongue "whose grammar displays such interesting varieties of syntax, gender, and expression" (Four-act, p. 52). Lady Bracknell prefers German vocal music to French, she says because unlike French "German sounds a thoroughly respectable language, and indeed, I believe is so" (p. 128). A truer reason for this avuncular preference may be Aunt Augusta's own none-too-latent Valkyrie tendencies. "On this point, as indeed on all points, I am firm" (p. 176)—"when my heart is touched, I become like granite. Nothing can move me" (Four-act, p. 160). In her capacity as aunt she is instantly recognizable ("Ah!" says Algernon, hearing the doorbell, "that must be Aunt Augusta") for announcing her presence in a "Wagnerian manner" (pp. 124–25).

Wagner is an important figure in many cultural systems by 1895, but not least in the systems of sexual signification. Himself certifiable as hetero-sexually active, if not hyperactive, Wagner nonetheless crystallized a hyper-saturated solution of what were and were becoming homosexual signifiers. This was possible partly because the newly crystallizing German state was itself more densely innervated than any other site with the newly insistent, internally incoherent but increasingly foregrounded discourses of homo-sexual identity, recognition, prohibition, advocacy, demographic specifi-cation, and political controversy. Virtually all of the competing, conflict-ing figures for understanding same-sex desire—archaic ones and modern ones, medicalized and politicizing, those emphasizing pederastic relations, gender inversion, or "homo-" homosexuality—were coined and circulated in this period in the first place in German, and through German culture, medicine, and politics. (At a time after the Wilde scandal when the author's name and work were elsewhere in total eclipse, his popularity in Ger-many, appropriately, was undimmed—including the 1903 publication of a play called *Ernst Sein*.) The intensely sexualized and nationalized Wag-nerian opera, accordingly, set up under the notorious aegis of Ludwig II, represented a cultural lodestar for what Max Nordau, in *Degeneration*, refers to as "the abnormals"; the tireless taxonomist Krafft-Ebing quotes a homosexual patient who is "an enthusiastic partisan of Richard Wagner, for whom I have remarked a predilection in most of us [sufferers from "contrary-sexual-feeling"]; I find that this music accords so very much with our nature." Dorian Gray, too, an Englishman, hears in Wagner's music "a presentation of the tragedy of his own soul." And an 1899 questionnaire by the pioneering gay sexologist Magnus Hirschfeld, designed to help readers gauge the degree to which they themselves might be "At All An Uranian" (i.e., an invert), included the indexical question, "Are you particularly fond of Wagner?" [18] The "Wagnerian" "ring" of Aunt Augusta, thus, however it may seem to herself a guarantee of perfect sexual rectitude, invokes very different relations in the nephews who attend it as spectators, and who cir-culate in an urbane world where the very name "Wagner" is a node of gay recognition and attribution.[19]

The germaneness of the German in *Ernst Sein* isn't only through aunts,

18. Bartlett, *Who Was That Man?*, p. 72.

19. Indeed, the association of Wagner with unauthorized sexuality may have to do specifi-cally with the heightened eroticism and violence he attaches to the relations of aunt, uncle, sib: see *Die Walküre, Siegfried, Tristan und Isolde, Lohengrin*, for example.

however, but through brothers, the alibi of the *hermano*—indeed German *is* the alibi of Jack's *hermano* in his urban mode. The German text is what's left guarding his aunt/niece/ward Cecily in the Hertfordshire place where brother Ernest isn't—where "Uncle" Jack isn't when he goes to London to be (with) Ernest; it is what Ernest as Algernon interrupts when he escapes Jack and finds his way to Hertfordshire. Miss Prism chides Cecily that Jack "laid particular stress on your German, as he was leaving for town yesterday. Indeed, he always lays stress on your German when he is leaving for town" (Cecily: "Dear Uncle Jack is so very serious!") (p. 141). And when the brother alibi begins to unwind, the threat of *too much* éclaircissement, of revelations that won't fit in with the marital telos, is wafted on the now familiar east wind:

> GWENDOLEN: Mr. Worthing, what explanation can you offer to me for pretending to have a brother? Was it in order that you might have an opportunity of coming up to town *to see me* as often as possible?
> JACK: Can you doubt it, Miss Fairfax?
> GWENDOLEN: I have the gravest doubts upon the subject. But I intend to crush them. This is not the moment for German scepticism. (p. 174; emphasis added)

"German," then, as befits the thick, complex, incoherent sexual valences of the label, both enables and threatens (what Christopher Craft describes deconstructively as) the treatment of fraternal bonds in this play as pure alibi: brotherhood as Bunburying, the empty, disavowed position "beside themselves" into which men may move at will, becoming "Ernest in town and Jack in the country": the imaginary bond by which Algernon (masquerading as Ernest) may try to relate to Jack *as Jack* (also masquerading as Ernest) *relates to himself*. Thus, insofar as the name "brother" functions as alibi, as pure structure, it is homologous with a modern/deconstructive understanding of men's bonds with other men as "homo," as relations to an (always absent, lacking, elusive or eloping) Same, in the homogenizing heterosexist scientism of homo/hetero.

Abstract brotherhood, then, is seen as Bunburying; but the highly cathected noun and verb "Bunbury" itself, the nonfamilial alibi in the play, as distinct from the term "brother," alludes (as several critics note) to surreptitious sexuality in terms of a particular genital practice: anal sex as "burying in the bun." The practice wasn't Wilde's own, but the repeated reference to it does add to the specific gravity of the play's *resistance* to homo/hetero scientism: in this case, by appealing to the premodern (though by no means obsolete) understanding of sexual nonconformity in terms of *acts* (e.g., "sodomy"), rather than types (inverts, homosexuals) or even relations (pederastic).

> ALGERNON: A man who marries without knowing Bunbury has a very tedious time of it.
> JACK: That is nonsense. If I marry a charming girl like Gwendolen, and she is the only girl I ever saw in my life that I would marry, I certainly won't want to know Bunbury.
> ALGERNON: Then your wife will. (p. 124)

So little is this transgressive *practice* seen as tied to the modern homosexual *type*, that Wilde doesn't even confine its attribution to his own gender: the same Gwendolen who's dangerously attracted to effeminate men is also, and apparently *not* as part of the same rhetoric, ascribed an eager anal sexuality that nothing in turn-of-the century (and little enough in contemporary) sexological "common sense" could have made any space for.[20]

What Jack means in saying, when they turn out to *be* brothers, that Algy has "never ['till to-day'] behaved to me like a brother" is, of course, that they have been very close friends. Unlike the glitteringly implausible cross-gender marriages that dramatically, inevitably arrive to cement the glitteringly implausible cross-gender courtships, the much less foregrounded arrival of a literal sibling bond to cap the bond of male attachment makes it seem more unsettlingly that something may really be changing—something may be lost. That there is something to be lost. The effect of something to be lost in this relation, something at a different ontological level from the other relationships in the play, is clearly traceable to a grammatical effect. The prevailing grammatical mode of the play is, of course, pointed wit— its almost ascetic paring down to the bare surface of the inversions, epigrams, double entendres that Camille Paglia rightly describes as "smooth with Mannerist spareness," "spasms of delimitation," "attempts to defy the temporal character of speech."[21] Wilde's "greatest departure from the Res-

20. My argument here is, I hope obviously, not the same as Joel Fineman's when he writes: "For now, remembering their etymology, we may rechristen the autological as the autosexual, or rather, the homosexual, and we may equally revalue the heterological as the heterosexual. This leaves us with the psychoanalytic conclusion that the fundamental desire of the reader of literature is the desire of the homosexual for the heterosexual, or rather, substituting the appropriate figurative embodiments of these abstractions, the desire of the man to be sodomized by the woman. . . . This would also explain why the only word that ends up being naturally motivated in *The Importance of Being Earnest* is not *Earnest* but *Bunbury* itself . . ." (pp. 88–89).

21. Camille Paglia, "Oscar Wilde and the English Epicene," *Raritan* 4 (Winter 1985): 94. A somewhat different version of this essay also appears in Paglia's *Sexual Personae: Art and*

toration dramatists," Paglia points out, is to "detach[ ] the witticism from repartee, that is, from social relationship."[22] It isn't so much that nothing is spared as that nothing is to spare—nothing doesn't pay off; almost literally, not a word of the play fails to contribute its full quantum to the clockwork mechanisms of syntactic and semantic parsimony and hyper-salience.

The exception to this law of salience is, then, precisely subliminal in its operation, a pattern of dull spots, single-entendre phrases, unnecessary words. They occur in the middle of the utterances Algernon and Jack address to each other. Here are some from one otherwise spectacularly quotable scene:

> My dear Ernest.
> My dear fellow.
> Dear Algy.
> My dear fellow.
> My dear fellow.
> My dear fellow.
> Old boy.
> My dear fellow.
> Dear boy.
> My dear fellow.
> My dear Algy.
> My dear Algy.
> My dear fellow.
> My dear young friend.
> My dear fellow.
> (pp. 117–24)

The most commonplace list of interpellatory endearments; murmurs from where?—clubland, the fifth-form study, the darkened bedroom; the earth tones of *The Importance of Being Earnest*. Neither figures of speech nor puns nor even constative propositions, these phrases, purely phatic, raising no issues of the "proper" or "improper" name, mean nothing *but* relation—do nothing but situate in an intimacy of two. In a play in which every relationship (including several that don't exist) is named and explicitly accounted for, sometimes several times over in conflicting versions, the unspecified bond of Jack to Algernon is taken as requiring no account. For

---

Decadence from Nefertiti to Dickinson (New Haven, Conn.: Yale University Press, 1990), pp. 531–71, as chap. 21, "The English Epicene: Wilde's *Importance of Being Earnest*."
22. Paglia, "Oscar Wilde and the English Epicene," p. 93.

the play's purposes, it *is* the natural. At least, it is so until it is *naturalized in the name of the Family*—when it becomes the stressful, invidious thing at which Algy can only strain at doing his best, "though I was out of practice."[23] If, as Algernon says in the four-act version of the play, relations "are a sort of aggravated form of the public" (p. 34), the main aggravation seems to consist in how the public grows *less* individual, less differentiated, even less lovable as it is brought within the holy bounds of family privacy.

Supposing we wanted to ask whether the play, as a play, narrows or extends, "stabilizes" or "destabilizes," the holy name of the family as our culture hands it to us: we would have to ask conclusively, at this point, a difficult question: what it means that the play's central marriage, the one between Jack and Gwendolen, can't take place until Jack is demonstrated to be, not only Algernon's "true" brother, but (as Ernest) his *own* "true" brother; Aunt Augusta's "true" nephew; at once Algernon's big nephew and big uncle (as Algernon is also marrying his "little aunt" cum "little niece" Cecily)—and, finally, his own wife's first cousin, mediator between the sibship and the avunculate, in the chiasmic, diagonal relation that in most cultures even now forms the immediate defining demarcation, from one side or the other, of the boundary legally called "incest": that between inside and outside the family.

To pose the question in this way, however, would be, not only to frame the play yet again in terms of its conclusion, but to enlist this entire reading in the service of two projects about which I think we now have the ability to make some more critical interrogations. I borrow a term from Leo Bersani in saying that what's limiting about both projects is that they are *redemptive*—that they presume and hence reinforce the essentially theological assumption that any cultural manifestation under study must respond first and last to the moralistic questions, "Can it be Saved?" and "Can it save *us?*"[24] In the political register it seems that the question toward which this reading might tend is, *Can the family be redeemed?* In the literary register

23. Interestingly, the introduction of the naturalizing name of the father in this play refers *not* to his surname, the legitimating aegis supposed to extend vertically to every child of his, but rather to the accident of his given name, which depends on the contingent seriality of the sib relation ("Being the eldest son you were naturally christened after your father" [p. 188]). If the paternal name is something you get from Canon Chasuble, *everyone* can be Ernest; if you have to get it from your father, only the *erst* can *Ernst sein*. Algernon, in a sense, loses his father by becoming a brother; loses his father as Jack wins him, at the same time winning Jack in the more indisseverable calculus of fraternity.

24. Leo Bersani, *The Culture of Redemption* (Cambridge, Mass.: Harvard University Press, 1990).

it would be, *Can the play be redeemed?* Both of these questions deserve, I think, to be resisted.

Can the family be redeemed? The easiest path of argument from some of my starting points here would be advocacy of a more elastic, inclusive definition of "family," beginning with a relegitimation of the avunculate: an advocacy that would appeal backward to precapitalist models of kinship organization, or the supposed early-capitalist extended family, in order to project into the future a vision of "family" elastic enough to do justice to the depth and sometimes durability of nonmarital and/or nonprocreative bonds, same-sex bonds, nondyadic bonds, bonds not defined by genitality, "step"-bonds, adult sibling bonds, nonbiological bonds across generations, etc. At the same time, as we have seen, a different angle—perhaps an avuncular angle—onto the family of the *present* can show this heterosexist structure always already awash with homosexual energies and potentials, even with lesbian and gay persons, whose making-visible might then require only an adjustment of the interrogatory optic, the bringing *to* the family structure of the pressure of our different claims, our different needs. Clearly, there is a very great deal that is imaginatively exciting and politically important in such programs: beginning but not ending with their affordances for lesbian and gay filiation and survival. Such programs are right at the center of a lot of present contestations around gay/lesbian issues, in terms of broadened definitions of domestic partnership; moves to pluralize what counts as "family" for legal, welfare, insurance, and real estate purposes; debates over whether a politics called "pro-family" need somehow *necessarily* be antichoice, antiwoman, antichild, and antigay; and, maybe most tellingly, the Thatcher government's framing of Clause 28, its charter of censorship against gay-affirmative speech, as a rebuttal of queers who present ourselves as what the legislation terms "pretended families." In the United States, I suppose the best definition of family is who you spend Thanksgiving with. The people I like to spend Thanksgiving with, joined to each other by being friends, lovers, companions, sometimes parents and children, and people who like to spend Thanksgiving together, proudly call ourselves after Clause 28: the Extended Pretended Family.

At the same time it seems that too much, too important ground is given up in letting the problematic of "family" define these intimate and political structurations—"family" even in the most denaturalized and denaturalizing, the most utopian possible uses of the term. One gay scholar/activist, Michael Lynch, said in a 1989 interview that he finds "family" a "dangerous word": "I don't like the idea of the gay family, it's a heterosexist notion.

I'd like a straight family to see themselves in terms of friends. I'd rather see same-sex friendship be the model to straights."[25] The very difficulty of conceiving this, a difficulty which the whole plot of *Earnest* underlines, points to the worst danger about "family": how much the word, the name, the *signifier* "family" is already installed unbudgeably at the center of a cultural value system—so much so that a rearrangement or reassignment of its *signifieds* need have no effect whatever on its rhetorical or ideological effects. You will have noted a certain impatience, in this reading of *Earnest,* with the concept of the Name of the Father. That is partly because I see what may have been the precapitalist or early-capitalist functions of the Name of the Father as having been substantially superseded, in a process accelerating over the last century and a half, very specifically by what might be termed the Name of the Family—that is, the name Family. (Within this family, the position of any father is by no means a given; there are important purposes, including feminist ones, for which the term "familialism" may now usefully be substituted for "patriarchy.") Now, the potency of any signifier is proven and increased, over and over, by how visibly and spectacularly it *fails* to be adequated by the various signifieds over which it is nonetheless seen to hold sway. So the gaping fit between on the one hand the Name of the Family, and on the other the quite varied groupings gathered *in* that name, can only add to the numinous prestige of a term whose origins, histories, and uses may have little in common with our own recognizable needs. Redeeming the family isn't, finally, an option but a compulsion; the question would be how to *stop* redeeming the family. How, as well, to stop being complicit in the process by which the name Family occludes the actual extant relations—for many people, horrifyingly impoverished ones; for everyone, radically changed and unaccounted-for, indeed highly phantasmatic ones—that mediate exchanges between the order of the individual and the order of capital, "information," and the state.

25. Nick Sherman, "Fighting Words" (interview with Michael Lynch), *Xtra!* 138 (8 December 1989): 7.

"Tales of the Avunculate"—written in the summer of 1990—was sparked by the work of Craig Owens, cited in the essay; and any pleasure in its writing came from the anticipation of showing it to him when a draft of it might be completed. That was the least of the things that suddenly became impossible on his death from AIDS-related illness, on 4 July 1990, at age thirty-nine.

David Shengold kindly advised me about Wagner.

IS THE RECTUM STRAIGHT?

IDENTIFICATION AND

IDENTITY IN *THE*

WINGS OF THE DOVE

"The Jamesian phantasmatic," Kaja Silverman remarks in a 1988 essay on subjectivity in Henry James, "can . . . be said to enclose homosexuality within heterosexuality, and heterosexuality within homosexuality."[1] Silverman's formulation—describing a mutual interiority between heterosexual and homosexual desires and identifications—would seem to crystallize a variety of de-essentializing currents in recent gender theory. That "homosexual" and "heterosexual" cannot be assumed to designate naturally distinct species, distinct affective universes, distinct populations, distinct forms of social and psychic organization is an implication of much work of the last half-decade in history, philosophy, anthropology, psychoanalysis, cultural studies, and literary criticism. Yet models for the mutual interimplication of "homosexuality" and "heterosexuality" remain fragmentary. Perhaps, for that matter, the fragmentary is their necessary or their most becoming form. They may not be fragmentary *enough*. Even summary statements like the one I have quoted from Silverman can suggest the risks—of reification, of historical decontextualization, of spurious symmetry—involved in the development of hypotheses on this subject. The problem is only exacerbated by the fact that psychoanalysis, profoundly as it has been shaped by homophobic and heterosexist assumptions and histories, has nevertheless not become dispensable as an

1. Kaja Silverman, "Too Early/Too Late: Subjectivity and the Primal Scene in Henry James," *Novel* 21 (Winter/Spring 1988): 165. A version of this essay also appears as chap. 4 in Silverman's *Male Subjectivity at the Margins*. Further cited in the text as KS.

interpretive tool for any project involving sexual representation. Thus, for example, Silverman's valuable reading of James, although adventurous and inventive in its most visible deployments of Freudian texts and formulations, seems to enjoy that privilege of invention only on the ground of several almost abject collapses of her argument into the most damagingly stereotyped of Freudian presuppositions. Silverman's is work that, among its other aims, seems to struggle to create a space within psychoanalytic theory for nonprejudicial, indeed appreciative interpretations of homosexual desires and orientations. Yet I would summarize the damages in her essay, for example, by saying that the transferential relations surrounding the analyst/critic—indeed, that *any* relations surrounding that figure—remain rigidly unexamined there; that, concomitantly, the understanding of James's literary production relies on a severely reductive, indeed a rather insulting model of repression and the unconscious; that it renders the formal and stylistic agency of James's texts invisible; that it excludes, or rather repels consideration of, every historical dimension involving power, oppression, and the consolidation or resistance of marked identities; and that its explorations of gay possibility occur exclusively within a framework (that of the "primal scene" and of the "positive" and "negative" Oedipal complex) whose structuration already, tacitly installs the procreative monogamous heterosexual couple as the origin, telos, and norm of sexuality as a whole. Thus it might have been truer to her own account of James's phantasmatic had Silverman offered a less elegantly symmetrical description of it; the psychoanalytic presuppositions of her reading practice suggest that if homosexuality can be seen as enclosed (imprisoned?) within heterosexuality, what encloses (cradles?) heterosexuality in turn remains a system of heterosexist presumption.

Yet I would repeat my judgment that the readings so arrived at are valuable and even necessary ones. Psychoanalytic thought, damaged at its origin, remains virtually the only heuristic available to Western interpreters for unfolding sexual meanings—meanings whose form and content, for that matter, are in turn likely to have been structured by the intuitions of psychoanalysis or of its precursor thought systems. In this essay I want to continue to make use of Silverman's article with and against a major James novel she does not discuss there, *The Wings of the Dove:* a novel whose subject, I will be arguing, is precisely the difference made by damaged origins— origins damaged by homophobia. With and against Silverman, I want to give examples of some interimplications of heterosexual and homosexual emergences: not abstractly symmetrical, but historical and highly contingent ones. I will be using *The Wings of the Dove* to argue, contra Freud, that the foundational presence of "a man" in a woman's coming-to-identity, or of "a woman" in a man's, need have no predetermined relation to hetero-sexist teleologies—given the likelihood that the internalized "man" or "woman" may themselves function as embodiments of same-sex desires or of queer identities. By tracing intersections like these, and by putting them in the context of James's distinctive erotic stylistics, I mean to offer one model for nonseparatist and at the same time antiheterosexist interpretations of sexual identity: interpretations that do not understate the perceptual and organizing power of particular, emerging sexual identities, but that do equal justice to the complex operations by which those identities reach out to and reorganize the surrounding ecologies of sexuality and of gender.

The character in *The Wings of the Dove* whom it would be safest to call a homosexual is presented shorn of any of James's celebrated subjectivity effects; he may indeed represent a man without a subjectivity. Yet he marks the origin both of the novel itself, in whose first scene he appears, and of the novel's most dynamic character, Kate Croy, whose father he is. Lionel Croy's homosexuality is spelled out in a simple code with deep historical roots: the code of *illum crimen horribile quod non nominandum est*, of "the crime not to be named among Christian men" and "the love that dare not speak its name." The code is, in short, that of naming something "unspeakable" as a way of denoting (without describing) male same-sex activity. In other contexts, such as "The Beast in the Jungle" (written just before *Wings of the Dove*), James searchingly and flexuously explored the limits of the unexpected rhetorical possibilities offered by this strange family of locutions, which specify the homosexual secret by failing to specify anything, speak by refusing to utter, and form a crux between starkly alternative, semantically "full" and modernistically "empty" readings of a single linguistic figure.[2] By contrast, the code of preterition around Lionel Croy functions flatly, crudely, almost comically. "What was it," Merton Densher at one point wants to know, "to speak plainly, that Mr. Croy had originally done?"[3]

2. This is treated much more fully in "The Beast in the Closet: James and the Writing of Homosexual Panic," which appears as chap. 5 of my *Epistemology of the Closet;* see esp. pp.201–3.

3. He asks it because "it was, naturally, . . . the question of her father's character that engaged him most"—and note how the repeated invocation of the "natural" in this staging or off-staging of the *contra naturam* is one of the things that comes closest to marking the discussion as comic. Henry James, *The Wings of the Dove*, ed. John Bayley (London: Penguin Books, 1986), pp. 98–99. Further citations of this edition will be given by page numbers incorporated in the text.

"I don't know [Kate responds]—and I don't want to. I only know that years and years ago—when I was about fifteen—something or other happened that made him impossible. I mean impossible for the world at large first, and then, little by little, for mother. We of course didn't know it at the time . . . but we knew it later; and it was, oddly enough, my sister who first made out that he had done something. . . . I suddenly heard her say, out of the fog, which was in the room, and apropos of nothing: 'Papa has done something wicked.' And the curious thing was that I believed it on the spot and have believed it ever since, though she could tell me nothing more—neither what was the wickedness, nor how she knew, nor what would happen to him, nor anything else about it. . . . I took her word for it—it seemed somehow so natural. We were not, however, to ask mother—which made it more natural still, and I said never a word. But mother, strangely enough, spoke of it to me, in time, of her own accord—this was very much later on. . . . She came out as abruptly as Marian had done: 'If you hear anything against your father—anything I mean except that he's odious and vile—remember it's perfectly false.' That was the way I knew it was true. . . . As it happens, however, . . . I've never had occasion [to contradict such an assertion] . . . . No one has so much as breathed to me. That has been a part of the silence, the silence that surrounds him, the silence that, for the world, has washed him out. He doesn't exist for people. And yet I'm as sure as ever. In fact, though I know no more than I did then, I'm more sure. And that," she wound up, "is what I sit here and tell you about my own father. If you don't call it a proof of confidence I don't know what will satisfy you."

"It satisfies me beautifully," Densher returned, "but it doesn't, my dear child, very greatly enlighten me. You don't, you know, really tell me anything. . . . What has he done, if no one can name it?"

"He has done everything."

"Oh—everything! Everything's nothing."

"Well then," said Kate, "he has done some particular thing. It's known—only, thank God, not to us." (pp. 98–99)

In a novel where the supposed glory of the unspoken things that continually "hang fire" seems stretched to the fineness of sublimity, it's notable by contrast that Lionel Croy's unspeakability—the novel's first—is kept by his family at the clownishly, squalidly formulaic level of epithets and invocations: his "indescribable arts" (p. 58); "whatever it was that was horrid—thank God they didn't really know!—that he had done" (p. 60);

the "much [Kate] neither knew nor dreamed of" about him (p. 61); "what they called the 'unspeakable' in him" (p. 64); his being, his sister-in-law says, "too bad almost to name" (p. 480); Kate's injunction—the last mention of Lionel in the novel—"If you love me—now—don't ask me about father" (p. 495). The implication is as pat as when an E. M. Forster character refers to himself as "an unspeakable of the Oscar Wilde sort."[4]

For all the demeaning flatness of the register of language that defines him, however, the "unspeakable" Lionel Croy is at the very source of the novel's energies. Comparing his fragmentary, framing presence in the completed novel with the "poor beautiful dazzling, damning apparition that he was to have been," James in the 1907 Preface to the New York Edition says that in his original plans for *The Wings of the Dove,* "the image of [Kate's] so compromised and compromising father was all effectively to have pervaded her life, was in a certain particular way to have tampered with her spring" (p. 43). James writes as if the novel as written didn't convey exactly this, but I think it does. Kate herself asks, "How can such a thing as that not be the great thing in one's life?" (p. 99). It seems to be so in hers, but seeing how depends also on seeing how Kate in turn, however "tampered with," or indeed because she has been "tampered with," herself remains the "spring" of the novel as a whole. Kate Croy's construction as a woman, her sexing and gendering, have her father's homosexual disgrace installed at their very core. That is to say, they have installed at their core the irresolvable compound of a homophobic prohibition with a nascent homosexual identity, both in a gender not her own. They are also intimately structured by an economic poverty that may or may not be the

4. E. M. Forster, *Maurice* (New York: W. W. Norton, 1981), p. 159. Specifying Lionel Croy as not just an unspeakable, but an unspeakable "of the Oscar Wilde sort," may be more than a figure of speech: I take Lionel in fact to be modeled on Wilde in some important respects. In 1895, James had written in a letter to Edmund Gosse, in response to Wilde's arrest, "it has been, it is, hideously, atrociously dramatic & really interesting—so far as one can say that of a thing of which the interest is qualified by such a sickening horribility. It is the squalid gratuitousness of it all—of the mere exposure—that blurs the spectacle. But the *fall*—from nearly 20 years of a really unique kind of "brilliant" conspicuity (wit, "art," conversation—"one of our 2 or 3 dramatists &c.,") to that sordid prison-cell & this gulf of obscenity over which the ghoulish public hangs & gloats—it is beyond any utterance of irony or any pang of compassion! He was never in the smallest degree interesting to me—but this hideous human history has made him so—in a manner." Rayburn S. Moore, ed., *Selected Letters of Henry James to Edmund Gosse 1882–1915: A Literary Friendship* (Baton Rouge: Louisiana State University Press, 1988), p. 126. Among the connections between the two figures may be Lionel Croy's strange position in the novel—soliciting "interest" even as he so decisively repels it, with a kind of sublimity of anticathectic abjection.

result of the father's homosexual disgrace. In turn, Kate's way of being sexed and gendered seems to be, in an important sense, what propagates gender and sexuality across the rest of the novel and across its characters.

I would like to emphasize how distant we already are, with these assertions, from a psychoanalytic understanding of the "primal scene" and the Oedipal complex as the main pathways through which gender and sexuality are constructed. In the first place, each of those models requires the assumption that individual gender and sexuality, whatever may be their variations, are arrived at by some, more or less complex routing of a child's identifications and desires through the circumscribed sexual dyad of a father and a mother. Kaja Silverman, for example, responding to James's insistent reproduction of sexualized tableaux and to his evident interest in anal positionalities and thematics, associates him with Freud's patient the "Wolfman," about whom Freud deduced that his patient's primal scene phantasy revolved around a glimpse of *coitus a tergo* between his parents (KS, pp. 162–66); Silverman, correspondingly, argues that "the Jamesian phantasmatic" sustains two desires: "the desire to be sodomized by the 'father' while occupying the place of the 'mother,' and the desire to sodomize him while he is penetrating the 'mother'" (p. 165). "Father" and "mother" are in scare quotes throughout Silverman's essay, but obviously the point of her analysis depends at the very least on identifying these *positions*, if not the persons actually occupying them, with one male and one female adult in an enclosed, procreative heterosexual bond. The identifications and desires that mark Kate Croy's sexuality, by contrast, though they do indeed involve both her father and her mother, cannot be arrived at through any permutation of the hermetically closed procreative heterosexual dyad. The formative fact of her father's homosexual disgrace requires, simply, obviously, the fact of her father's sexual engagement with someone not her mother and not a woman. It is valuable that such a constellation should be irrecuperable in terms of the Oedipal configuration. For, while there is immense heuristic value in psychoanalysis' formalistic requirement that every sexual variation be shown to be somehow derivable from the cross-gender dyad (the very difficulty and unlikelihood of the task generating ever subtler, more flexible explanatory tools for derivation), nonetheless there are very substantial and unacknowledged costs attached to the theoretical parsimony of the Oedipal scenario. The circularity of the connection between heterosexist presumption and gender definition itself, gender as founded in the Oedipal scenario, has been sufficiently demonstrated by work like that of Judith Butler. So it could make a difference, to an unpredictable degree, to insert homosexuality into this framework not in the position of a possible *outcome,* but rather in that of an available—I will be arguing *routinely* available—tutelary *model.*

The potential importance of that move may not depend, either, on the ethical valuation assigned the homosexual "model"—in fact, as far as *Wings of the Dove* goes, it had better not. No role model, no ego ideal, Lionel Croy presents the most demeaned and demeaning figure imaginable in the novel. The interpretive paths by which there is any sense to be made of him are completely paved, as I'm afraid I may be unable to stop demonstrating, with the *idées reçues* of homophobic "worldly wisdom." Indeed, it is *only* the world's word that constitutes his homosexuality and himself; he lacks not only the dignity of a subjectivity, but even the dignity of any actual desires. Lionel attributes this to his discursive isolation; when Kate, giving up her attempt to evoke "normal" paternal feelings from him, muses, "I don't know what you're like," he responds, "No more do I, my dear. I've spent my life in trying in vain to discover. Like nothing—more's the pity. If there had been many of us and we could have found each other out there's no knowing what we mightn't have done" (pp. 66–67). Yet even his subjective vacancy adds to the exemplary resistance he poses to the psychoanalytic reader—exemplifying, I mean, the sense in which that modern invention "the homosexual" embodies less a set of subjective desires than a particular, historically contingent process of labeling applied in the first place from without. There are plenty of same-sex desires in *The Wings of the Dove;* but Lionel Croy, the character who bears, if not their name, then their far more pointed namelessness, is not seen to experience them. The shaping pressure he brings to bear on his world does not transform it in the image of his specific queer desire but rather in the image of his specific queer stigmatization—"the force of his particular type." His most striking legacy to Kate, for example, seems to be manifest in her manipulative, creative, desire-inducing genius for leaving brutal things unsaid: a trait that must be modeled, not precisely on himself, but rather on the assertive, brutalizing silence that surrounds his presence in the world as a homosexual. And in another of the novel's many examples of queer tutelage, what Susan Stringham has to teach Merton Densher at the novel's end, despite the *relatively* speakable (if demeaned) history of her own love for Milly and the obsolescent but legitimated proto-lesbian identity she has been able to occupy in relation to it—the supposedly saving lesson she is able to imprint on Merton at a climactic point is, again, the lesson of silence, the mendacious, intensely performative silence whose history is that of the closet.

I want to follow Judith Butler in suggesting that the gender and sexuality so instituted in the novel's world are profoundly shaped by the melancholia—the denied mourning—caused by these originary foreclosures of (a relatively specified) homosexual possibility.[5] I want at the same time to emphasize how little this process can be understood without giving an ontological gravity quite denied by psychoanalysis to the status of historically contingent identities. Aside from the special status granted by psychoanalysis, especially under the influence of Lacan, to the identities "male" and "female," all other complex intersections of behavior, subjectivity, self-perceived identity and other-ascribed identity are treated as both completely transparent and historyless by psychoanalytic discourse. This is disastrous for any project that does not aim simply to presume and reinscribe the lie of universal heterosexuality—given the diachronic shiftiness and ideological/epistemological impactedness of non-heterosexual (and by consequence, indeed, of heterosexual) identity formations since at least the mid-nineteenth century. For example, Silverman appears to use "sodomy" and "pederasty" almost interchangeably as supposedly value-neutral names for male-male anal penetration, regardless of their histories, their associations, even their denotations: regardless of the biblical derivation of "sodomy" from the name of a site of homophobic genocide, regardless of the historical plurality of the kinds of acts denoted by "sodomy" at different times and places (including *every* form of non-procreative sex with woman, man, or beast), regardless of the fact that the word's current usages are exclusively legal or theological, and uniformly interdictive; and regardless of the fact that "pederasty" is not the name of a genital act at all but of a historically specific, relational orientation of desire by an adult toward a youth. (Silverman refers, for instance, to a child's "pederastic" desire to sodomize its father! (p. 165)) This proceeding assumes that desires, behaviors, and identities can be translated unproblematically back and forth, one to another; worse, it assumes that neither the spectacularly differential levels of proscription against different sexualities, nor the distinct and changing *structures* of their proscription or embodiment, are of any consequence for psychic life.[6]

"The force" of Lionel's "particular type," as an unspeakable of the Oscar Wilde sort, isn't, as we have noted, visible in any actual homoerotic desires or acts—though it is worth wondering how much that aporia may result from the novel's seeing him only through the eyes of a daughter who

5. Butler, *Gender Trouble*, pp. 57–72.
6. In *Male Subjectivity in the Margins* Silverman revises this construction, changing "pederastic identification" to "sodomitical identification" (pp. 174, 179–81). This revision, however, still does not adequately address the historical and political complexities of sodomy, gender and homosexuality and their relationship (often tenuous or nonexistent) to each other.

(miming and perhaps misunderstanding her mother) repeatedly, formulaically repudiates the knowledge of "whatever it was . . . that he had done." How, instead, is his "type" in all its influence on the novel to be located? First, as I have already suggested, in a contagious sexual aesthetic (and a related novelistic aesthetic) that places a high value on brutalizing silences. Second, in the circulating "symptom" of sexual indifference. And third, in a distinctively connoisseurial, uncannily bifurcated and bifurcating visual relation to female beauty.

Teresa de Lauretis popularized the term "sexual indifference" in an influential 1988 article called "Sexual Indifference and Lesbian Representation," where, silently overriding the vernacular meaning of the word "indifference," she instead uses it to denote the problematic of sexual *undifferentiation,* lack of gender difference, in male homosexuality and lesbianism.[7] Although an aim of her argument is to complicate the heterosexist psychoanalytic assumption that gender difference is the only imaginable basis for sexual attraction, it remains true in this essay that the compulsive, pointless English-French punning that makes "indifference" (a word with a perfectly good meaning of its own) dissolve so neatly into "lack of difference," as if the two meanings were after all the same, finally re-performs, at the verbal level, exactly the heterosexist work of those psychoanalytic assumptions. But just as the absence of gender difference is by no means a guarantee of anybody's sexual indifference (au contraire), by the same token sexual indifference itself is by no means a transparent phenomenon whose meaning can be taken for granted, or seamlessly subsumed under the label of some supposed explanation. Libidinal indifference, as I think most of us could testify who have ever desired somebody who just didn't desire us, is a force in its own right. It changes lives. And it doesn't only operate in the exceptional case of the would-be love object who is, shall we say, the "wrong" gender for the lover: the plain fact is that most people in the world, whatever their gender or sexuality, don't form or maintain libidinal cathexes toward most other people in the world, whatever theirs. Not that they fail to, not that they can't, not that they do and then repress them, just that they don't: a plain but not especially simple fact; a negative space in theories of desire, but one that has a shaping impact, as well, on desire itself.

It is sexual indifference in this sense that marks Lionel Croy—and in marking him, marks the gender and sexuality of the daughter who has found no path into his affections. Kate, as we will see, takes her place in

7. Teresa de Lauretis, "Sexual Indifference and Lesbian Representation," *Theatre Journal* 40 (May 1988): 155–77.

Wings of the Dove as part of a long tradition of novelistic and filmic women who at the same time function as embodiments of sexuality itself, and yet are presented as "frigid." In the context of this particular novel, the latter of these traits clearly represents the impress of her father; the former may, as well. At the start of the novel she has gone to visit her father in all his diminished state, offering to throw in her lot with him and give up being the protégée of her wealthy Aunt Maud. Lionel, his own interests too slickly sordid for such sentimental renunciations, will have none of it. The scandal of his active, persistent, and very complete impermeability to the appeal of family feeling—to the appeal, that is, of the sacrifice Kate wants to make in order to act a daughter's part with him—deserves to be called *sexual* indifference, not just familial indifference, partly because it so takes its shape (for her) from (her perception of) the aggressive surface of refusal he has presented to her mother. "She now again felt, in the inevitability of the freedom he used with her, all the old ache, her mother's very own, that he couldn't touch you ever so lightly without setting up. No relation with him could be so short or so superficial as not to be somehow to your hurt" (p. 57). It is interesting that the language of sexual refusal here sounds so much like the language of sexual abuse. The *ache* of "the freedom he used with her" (itself a redolent phrase) would sound like the ache of a nonsexual, resistant response to sexual aggression—if it didn't, in context, sound even more like an eroticization of the place where a sexual incitement has failed of presenting itself. That the very superficial, surfacial structure of Lionel's indifference undermines the boundaries between self and other as piercingly as could desire itself is clear, at any rate, in the inside-out syntax of Kate's feeling "there was never a mistake *for you* that he could leave unmade, nor a conviction of his impossibility *in you* that he could approach you without strengthening" (p. 57; emphasis added). What we can see as having been implanted in Kate by such penetrations is a double response to her father's sexual indifference: sentimentally resisting it in him, she yet also reproduces it in her behavior toward others—but reproduces it, ache and all, with the surplus of her own energy and interest, *as* sexuality.

However subtly its effects on Kate may be psychologized, the identity between Lionel Croy's "unspeakability" and his frigidity represents a quite specific—a historically specific—homophobic libel, not the less so in that permission to imagine him as cathecting anyone *outside* his family (e.g., a man or boy) is foreclosed by the same gesture as is his desire for the women attached to him. He is worlds apart, for instance, from the indiscriminately and ungovernably appetitive figure that would have been a medieval or early modern "sodomite." Equally a homophobic libel, equally a staple of the turn-of-the-century "worldly" coding of gay male roles, is the specificity of the one way in which Lionel does engage with his daughter: through a critical, connoisseurial appreciation of her looks and clothes.

> He judged meanwhile her own appearance, as she knew she could always trust him to do, recognising, estimating, sometimes disapproving, what she wore, showing her the interest he continued to take in her. He might really take none at all, yet she virtually knew herself the creature in the world to whom he was least indifferent. She had often enough wondered what on earth, at the pass he had reached, could give him pleasure, and had come back on these occasions to that. It gave him pleasure that she was handsome, that she was in her way a tangible value. (p. 59)

Kate's appearance here is labeled "handsome" only a few sentences after she has called her father's by the differently gendered epithets "beautiful" and "lovely" ("You're beautiful . . . you look lovely"): sufficient suggestion, if it were needed, that the coding of gay male cultural work here (perhaps indebted to Oscar Wilde's tenure as editor of the magazine *Women's World*) is continuous with the still surviving folk "knowledge" about the misogyny of gay fashion arbiters who are driven by the (not necessarily compatible) imperatives to (a) make money off women, (b) make women suffer, (c) aestheticize women, (d) make women hideous, and (e) make women look boyish and therefore attractive to their own supposed sexual tastes.

Kate indeed is said to resemble "a breezy boy" (p. 165), and the effect of her "handsomeness," her only avenue of appeal to her father, is at any rate important enough in the novel to be worth pausing over. The reader's first view of it is Kate's view of herself in a mirror in the room where she is waiting for him; and she, like him, or indeed as if already marked by him, seems to appeal to her own beauty *first* as "a tangible value"—as "her nearest approach to an escape" from the hard questions set by poverty and inarticulable ruin.

> Wasn't it in fact the partial escape from this "worst" in which she was steeped to be able to make herself out again as agreeable to see? She stared into the tarnished glass too hard indeed to be staring at her beauty alone. She readjusted the poise of her black closely-feathered hat; retouched, beneath it, the thick fall of her dusky hair; kept her eyes aslant no less on her beautiful averted than on her beautiful presented oval. (p. 56)

A striking phrase, "her eyes aslant no less on her beautiful averted than on her beautiful presented oval." But, as my favorite Linda Barry character would say, what does it even mean? Is Kate looking at her butt in the mirror? Is she watching the back of her head? Or is it her face itself that has split to offer not one aspect but two—or rather, to offer one so that it may be seen to withhold the other? A moralistic formulation would be that the pressure of want and disgrace has made her two-faced. And that's not far from true, if you can subtract *its* punitive patness: the simple fact of *sidedness,* double-sidedness, presented and averted, recto and verso, seems not just to be lodged in Kate's person but to radiate out from it across the novel. As the interview with Lionel goes forward we realize that in looking at herself in this way, "too hard indeed to be staring at beauty alone," but critically, unerotically, and as it were objectively, cloven by her own "tangible value" under the pressure of need and disgrace, Kate is looking at her beauty, averted and presented, not *alone* but also, already, through the eyes of her father's imposed sexual identity and as divided by the bar of its prohibition.

Whatever insult Lionel has offered to the gender system, its effect on his daughter is evidently not to release her from the constraints of gender but rather to force her the more grindingly to articulate and embody them. Eyes "fixed" on the mirror, Kate muses that "had she only been a man" "it was the name, above all, she would take in hand—the precious name she so liked and that, in spite of the harm her wretched father had done it, wasn't yet past praying for. She loved it in fact the more tenderly for that bleeding wound. But what could a penniless girl do with it but let it go?" (p. 57). Kate's wish to have been a man, so as to salvage the authority of the paternal name, is haplessly interlined with her self-definition as a woman, as an object of exchange who can't properly *have* a name. And it is the "bleeding wound" of Lionel's homosexual disgrace—pointed or exacerbated by material want—that attaches her more intimately, more "tenderly" to the site of gender bifurcation. Not that she is presented as conventionally feminine in herself: ever the "breezy boy" she remains handsome, gallant, and "brutal" with "a wild beauty" (p. 171). But her habit of being, as she sees herself in her father's mirror, "somehow always in the line of the eye" (p. 56) seems to divide those who view her, as it divides her own visage, with the violence of gender distinction. One of the words that turns up most regularly in odd usages in Kate's vicinity is "different." Differing herself, from herself, Kate seems to drive home to people the point or the question of binary difference. Densher recognizes in her "a woman whose value would be in her differences" (p. 87); "You're different and different—and then you're different again," he tells her (p. 289). "The difference," after they have had sex, is what she leaves in her wake (p. 399). When they talk, "there *was* a difference in the air—even if none other than the supposedly usual difference in truth between man and woman" (p. 120). She makes Milly Theale wonder "why she [Milly] was so different from the handsome girl—which she didn't know, being merely able to feel it; or at any rate . . . why the handsome girl was so different from her" (p. 152); similarly she makes Merton wonder "if he were as different for her as she herself had immediately appeared" (p. 449). The "difference" in Kate's vicinity doesn't only mean gender difference (it can be the difference between money and want [e.g., p. 390], sex and no sex, truth and lie, then and now), but what Kate in her two-sidedness does most characteristically is to apply stress to the impossible question of where difference can be located—thus functioning, almost inevitably, as an agent for propagating the gender bifurcation she struggles, herself, to refuse.

The most obvious case of this is, of course, Milly Theale, who sees in Kate "a figure conditioned . . . by the great facts of aspect" (p. 190): strikingly, it is not (for Milly) Kate's own gender but the gender of Milly as her viewer that Kate's trait of "aspect," her sidedness, has the effect of fracturing. During the time when their mutual acquaintance with Merton Densher remains—with increasing pointedness—unmentioned between the two friends, Milly is subject to a series of what can only be called strange fits of aspect when she regards the boyish beauty. At first these fits are voluntary in Milly, something of a guilty, induced pleasure: "Milly, who had amusements of thought that were like the secrecies of a little girl playing with dolls when conventionally 'too big,' could almost settle to the game of what one would suppose [Kate], how one would place her, if one didn't know her." This game of free-floating identity becomes less and less free, more and more uncanny and determined, however, as Milly loses her control over the onset of these fits, and as it becomes clear to her that the Kate being viewed at these moments is *not* Kate as she might appear to a random stranger, but rather Kate as she "must" appear to a very particular other person, a man, a man in love with her, Merton Densher. In what is presented as a typical instance, Milly watches from her balcony as Kate gets out of a cab:

> What was . . . determined for her was, again, yet irrepressibly again, that the image presented to her, the splendid young woman who looked so particularly handsome in impatience, with the fine freedom of her signal, was the peculiar property of somebody else's vision, that this fine freedom in short was the fine freedom she showed Mr

Densher. Just so was how she looked to him, and just so was how
Milly was held by her—held as by the strange sense of seeing through
that distant person's eyes. It lasted, as usual, the strange sense, but
fifty seconds; yet in so lasting it produced an effect. (p. 220)

"Just so was how she looked to him, and just so was how Milly was held
by her." Milly is not virilized by her desiring regard of Kate, but it leaves
her vulnerable to being involuntarily assigned to a subject position she can
only label male. Unable at the same time to be herself and to be the person
to whom Kate looks "just so," Milly responds to the sight of Kate's splendor by experiencing her own visual faculty as demonically possessed by a
man and alienated from herself. Her vision splits into two gendered halves,
but only one of the two belongs to her. Through this uncanny division or
excision, so reminiscent of the scenes of syncopic visual fracture in lesbian
vampire fictions like "Christabel" or *Carmilla*, Milly's "self" is consolidated as female; her sense of her own desire, furthermore, becomes riveted,
not on the Kate at whom she casts these admiring glances, but rather on the
Merton Densher she imagines herself to be in casting them.[8] It is *because*
her sense of gender is so radically bifurcated by the sight of Kate, by Kate's
handsomeness and her divided and divisive "aspect"—traits that bear so
much the mark of Lionel Croy—that we must describe Milly's gender and
sexuality (her femaleness, her heterosexuality) as among the products of
Lionel Croy's preterited homosexual disgrace. I am also saying, a fortiori,
that they are the product of (and represent the melancholia of the foreclo-
sure of) her lesbian desire for Lionel Croy's daughter.

Of course, it takes a variety of economies and discourses to constitute
any person as gendered and sexed. What being a woman is for Kate has
everything to do with the sharpness of material need that is part of her
legacy from her father. By the same token, Milly's femaleness can only be
embodied in relation to the other great facts about her: she is very rich;
she is very sick. As Cyndi Lauper might put it, illness changes everything.
Every aspect of being a woman, for Milly, is a form of relation to her wealth
and her illness; she can't be a person gendered female without also being
a sick role gendered female and a rich role gendered female. A climax of
Milly's performance in the rich role, and what seems to be the founding
moment of her sick role, occur in the famous scene of her great social triumph at Matching, a country house where, herself the cynosure of eyes, she

8. For writing on lesbian vampirism, see for example Sue-Ellen Case, "Tracking the Vampire," *differences: A Journal of Feminist Cultural Studies* 3 (1991): 1–20, and Paglia, *Sexual
Personae*, pp. 317–46.

is brought face-to-face with a Bronzino portrait that everyone perceives as resembling her, "with her slightly Michaelangelesque squareness, her eyes of other days, her full lips, her long neck, her recorded jewels, her brocaded and wasted reds. . . . And she was dead, dead, dead" (p. 196). Milly finds herself, "for the first moment, looking at the mysterious portrait through tears"; and the development of this extraordinary passage is the first place where it becomes palpable in the novel, and through Milly's point of view, that she is indeed gravely ill . . .

> and Kate was saying to her that she hoped she wasn't ill.
> Thus it was that, aloft there in the great gilded historic chamber and the presence of the pale personage on the wall, whose eyes all the while seemed engaged with her own, she found herself suddenly sunk in something quite intimate and humble and to which these grandeurs were strange enough witnesses. It had come up, in the form in which she had had to accept it, all suddenly, and nothing about it, at the same time, was more marked than that she had in a manner plunged into it to escape from something else.

"It," Milly's first, complex and almost luxuriant plunge into the full experience, the full social identity of illness, though it would seem to be only the long-awaited revelation of a primary fact about her, is curiously described instead as being a secondary reaction *against* the consciousness of "something else" even more threatening, something that is described as primary: and that is the fact of Kate's sidedness and of the bi-gendered double vision it always seems to implant in Milly.

> Something else [the passage goes on], from her first vision of [Kate's] appearance three minutes before, had been present to her even through the call made by the others on her attention; something that was perversely *there*, she was more and more uncomfortably finding, . . . by some spring of its own, with every renewal of their meeting. "Is it the way she looks to *him?*" she asked herself—the perversity being how she kept in remembrance that Kate was known to him. (p. 199)

Milly's style of being sick, as it emerges in the second half of the novel, will be characterized by an almost violent repudiation of the pity of others. What seems present in this foundational scene is, again, the desire as against which that repudiation is destined to come into being: it attaches, again, to Kate, whose half-expression of pity fills Milly not with the anger she will later display but with "a yearning, shy but deep, to have her case put to her just as Kate was struck by it. . . . She took Kate up as if positively for the deeper taste of it" (p. 201). Similarly, when Milly asks Kate to accompany her on a secret visit to consult the great medical specialist Sir Luke Strett, her decision is a compensatory reaction against the cloven aspect that Kate presses on Milly's desiring vision. It is as though Milly had to invent a lack for herself to answer to the rift in Kate's self-presentation: "She had wanted to prove to herself that she didn't horribly blame her friend for any reserve; . . . could she show it more than by asking her help?" (p. 202). Even more, it is as though Milly must invent a secret for herself to correspond to the secret that has divided Kate. In these ways Milly's experience of her illness, a crucial part of the construction of her femininity, is presented as a secondary reaction-formation, the status of primary cause being accorded to her feelings about Kate.

Milly's refusal of pity—refusal even of empathy or communication—has another antecedent as well, however. One of the most astounding effects in *The Wings of the Dove* is the way Milly's relation to her illness winds up, rhetorically, as an echo precisely of Lionel Croy's homosexual disgrace. The strangeness of this is of course that Milly's illness is supposed to be the most sacralizing element of the novel, as Lionel's disgrace is the most sordid. Yet the same compulsive, reifying gestures of disavowal surround both secrets. Early on, Milly says about her illness, "The best is not to know. I don't—I don't know. Nothing about anything" (p. 317). Kate instructs Merton in how to think about Milly's illness: "Don't guess" (p. 285). Susie Stringham and Maud Lowder discuss it in terms like these:

> "She hasn't what she thought."
> "And what did she think?" Mrs Lowder demanded.
> "He [Sir Luke Strett] didn't tell me."
> "And you didn't ask?"
> "I asked nothing," said poor Susie . . . .
> "Examining her for what she supposed he finds something else?"
> "Something else."
> "And what does he find?"
> "Ah," Mrs Stringham cried, "God keep me from knowing!" (pp. 320–21)

And in the Preface to the New York Edition it is James in propria persona who invokes the same divine shield—with a hysterical repetition—against the same knowledge: "Heaven forbid, we say to ourselves during almost the whole Venetian climax, heaven forbid we should 'know' anything more of our ravaged sister than what Densher darkly pieces together" (p. 46). When Kate Croy and her family repeatedly "thank God they didn't really know" what Lionel's "unspeakability" was, the effect is hardly more flatly formulaic, and indeed hardly more prurient, than when the characters and author of the novel repeatedly and so relishingly repudiate a knowledge of Milly's disease. The effect, in fact, is so similar—the hushed sounding so like the hush-hush—as to generate a kind of pornography of illness around Milly. If anything, the preterited illness of "our ravaged sister" is notable for being *more* eroticized than the numbly preterited sexuality of the older man: unspeakable illness, unlike unspeakable sexuality, has room in it for a candid and unconventional desire like Milly's desire of Merton. Similarly eroticized is James's description of the secret of Milly's illness as "what Kate Croy pays, heroically, it must be owned, at the hour of her visit alone to Densher's lodging, for her superior handling and her dire profanation of" (p. 46): a description that makes Kate's having sex with Merton sound like a grim punishment, and her epistemological manipulation of Milly sound like sex.

One way of considering the strange echo between the "unspeakabilities" that frame the novel might be to see Milly's unspeakable illness as a wishful recasting of that original unspeakability: where Lionel is trundled offstage in a hurry, insulted in his presence and in his absence, denied access to either the tragic or the comic register (p. 58), and deprived of both agency and desire, Milly on the other hand is the object of all the narrative "tenderness" with which "an unspotted princess is ever dealt with" (p. 50) in spite of her open erotic demands on an unsuitable young man—and has also, it seems, the luxury of being the person who is seen to impose the awful silence that surrounds her, rather than being, like Lionel, ignominiously forced into submission to it. A thin luxury it may be to choose silence rather than have it enforced; thin especially when the price paid for it is to be defined as incurably ill. (And I am far from wanting to suggest that it is the most dignified way to deal even with incurable illness.) Yet the way Milly's access to both pathos and desire has been rendered unproblematical under the topos of unspeakability might well seem enviable to Lionel, the original referent of that topos; and might for that matter seem a kind of wish fulfillment to the author who framed the two characters in this revisionary relation to one another. Another way of seeing the strange congruence between Lionel's stigma and Milly's fatal illness, however, would be to understand Milly's extremity, which so fills the latter half of the book with its perfume of exhaustion, mourning, and hush, as dramatizing the licit, indeed culturally much-valued melancholia (call it "heterosexuality"?) that covers over but can't cure the raw wound of the excision of homosexual possibility.

Unspeakable like Lionel or incurable like Milly, criminal or pathological, vicious or pathetic—this was the cleft stick in which people found themselves trapped at the turn of the century when they were interested in articulating explicitly gay subjectivities, or ideologies that would defend anyone labeled homosexual against the grossest punishment. As Jeffrey Weeks, Wayne Koestenbaum, and others have shown, a turn away from the criminal model of homosexuality in this period seemed as if inevitably to involve a turn toward the medical model: from Lionel to Milly.[9] Where the development of an account of homosexual *agency* was not in question, however, there actually was a range of other nonheterosexual male "types" available for discursive circulation and allusion; and it is, interestingly, representatives of two of these types—call them the urban bachelor and the gondolier—who preside even more than does Susan Stringham as the main tutorial presences over Milly's self-formation, as a woman, at the intersection of her new illness and her inherited wealth. The first of these is her doctor, Sir Luke Strett, an apparently unmarried man (the only family role we see him in is, characteristically, the avuncular) who offers Milly an exciting, sustained, formative intimacy that is different from her other relations with gentlemen in never raising any issue that would entail "sexual" —marital—narratives or consequences. Like the unmarried metropolitan physicians, journalists, and artists in other turn-of-the-century fictions of Robert Louis Stevenson, James M. Barrie, George Du Maurier, Wilde, Proust, and James himself, Sir Luke personifies the bachelor professional, a type whose historical relation to the emergence of gay identities I've discussed at some length in *Epistemology of the Closet*.[10] Milly's other tutelary figure is her Venetian majordomo Eugenio, "polyglot and universal, very dear and very deep" (p. 335). It is this insinuating figure, who understands Milly as nobody yet had understood her (p. 335), whom in a 1986 essay Michael Moon persuasively assigns to "a 'type' of whom the recently . . . recovered social history of turn-of-the-century homosexuality has made us aware: the lower-class Italian man whose services to his upper-class [male] employer[s] often secretly (?) included sexual ones."[11] Moon points out that there was an intense " 'trade' in gondoliers" and gondolier

9. Among the now considerable literature on this dilemma, the first, influential formulation of it was in Jeffrey Weeks, *Coming Out: Homosexual Politics in Britain, from the Nineteenth Century to the Present* (London: Quartet Books, 1977), pp. 23–32; the best discussion of its functioning in terms of a particular subjectivity is in Wayne Koestenbaum, *Double Talk: The Erotics of Male Literary Collaboration* (New York: Routledge, 1989), pp. 43–67.

10. See my *Epistemology of the Closet* chap. 4, esp. pp. 189–95.

11. Michael Moon, "Sexuality and Visual Terrorism in *The Wings of the Dove*," *Criticism* 28 (Fall 1986): 440–41.

icons among James's own circle (such people as John Addington Symonds, Horatio Forbes Brown, and John Singer Sargent). The Anglo language surrounding these figures, even (or especially) when most erotically charged, is routinely racist—but no doubt it was also the sense of their "racial" distance that seemed to make them erotically available for discursive circulation. Although (as Moon points out) Eugenio, along with his "decorative" sidekick "brown Pasquale" the gondolier (p. 358), figures in a crescendo of cruisy visual encounters and climaxes involving other men (Sir Luke, Merton, Lord Mark) in the Venetian chapters of the novel, his duty toward Milly is not to offer sexual service but to "*inclusively* assist her" in manipulating her wealth to make a habitable frame for her illness. For this he seems to be peculiarly well-equipped, and in his capacity of presiding over Milly's self-construction as woman, rich, and ill, he is explicitly compared to Sir Luke (p. 341). Less explicitly—as a sinister father, a de(hetero)sexualized father, a smoothly plausible and mercenary father, but in this case somehow a *good* father—he seems also to offer a gloss on the "beautiful" (p. 59) unspeakable father Lionel Croy, whose "plausibility had always been the heaviest of [Kate's] mother's crosses" (p. 60). James says of Eugenio,

> Gracefully, respectfully, consummately enough—always with hands in position and the look, in his thick neat white hair, smooth fat face and black professional, almost theatrical eyes, as of some famous tenor grown too old to make love, but with an art still to make money—did he on occasion convey to [Milly] that she was, of all the clients of his glorious career, the one in whom his interest was most personal and paternal. (pp. 335–36)

If Kate and Milly, in their different ways, can be shown to be gendered and sexed in the course of *The Wings of the Dove* under queer tutelage, they are not the only ones, and indeed that is not their only role in the tutelary economy. The homoerotic current that runs between the two women themselves is in some ways the opposite, ontologically, of Lionel Croy's homosexuality—his is all stigmatized identity in the absence of any desiring relation, while theirs subsists as a highly mediated circuit of desire in the abeyance of any homosexual labeling process.[12] Their homoerotic bond

12. Abeyance, I'd suggest, rather than absence, because Susan Stringham's presence in the novel (like, for that matter, the presence of *The Bostonians* in the historical background of James's later novels) offers a glimpse of an earlier generation of New England women for whom same-sex devotion was indeed, and directly, identity-constituting. The lapse of such possibilities at the turn of the century is dramatized when even Susan Stringham tries to in-

is, nonetheless, powerfully formative of the identities of others. The male gender and the heterosexuality of Merton Densher, for example, must be seen as being formed, not just in relation to Kate and hence Kate's father, but more immediately in relation to Kate's intense though comparatively inexplicit bonds with women: with Maud Lowder as well as with Milly Theale. Densher deprecates his situation in "a circle of petticoats" (p. 384); well may he feel that the fine freedom with which Kate turns her wealthy women friends into marital aids for herself and Densher is inefficient compared to the yet finer with which she turns her fiancé into an instrument for giving pleasure to her women friends. For all Kate's sharp decisiveness and executive energy, she uses her high abilities to postpone, always, instead of expediting—where postponement means keeping her place in the circle of petticoats, the circuit of mediated desires involving Merton, Milly, and Maud together, while expediting would mean sealing herself off from the other women in a dyadic heterosexual clinch. Bred to visibility, Kate has a keen appetite for Densher's "long looks" (p. 92) but is in no apparent hurry to go to bed with him and evinces no pleasure whatever when she has done so. But isn't it just her complete lack of interest in sex with him, compared to the fierceness of her focus on maintaining the triangular choreographies involving Maud and Milly, that gives Merton, in his frustration, the acutest sense he ever has of himself as man, as a heterosexual, as—once he has bargained her into "coming to his rooms" in Venice—master? He also feels later that the same bargain has defined him as a slave: the humiliated puppet of Kate's obsession with Milly. Masculinity as Densher inhabits it is a roller-coaster drama of sadism and masochism, and both come into being as he negotiates his positional possibilities in relation to Kate's unflagging intentness on other women.

Kate's double-sidedness, for example, at one clinching moment of their courtship, appears to him as the almost sickeningly arousing "discipline" of her face as she is forced to serve the theatrical needs of both her wealthy patroness and himself: at a party at Maud Lowder's, the relation between Kate and Maud seems

> a relation lighted for him by the straight look, not exactly loving nor lingering, yet searching and soft, that, on the part of their hostess, the girl had to reckon with as she advanced. It took her in from head to foot, and in doing so it told a story that made poor Densher again the

terpret young Milly through the pathologizing, ultimately heterosexualizing grid of neurosis, as a "case" of "American intensity" prone to "some complicated drama of nerves" (p. 130).

least bit sick: it marked so something with which Kate habitually and consummately reckoned. . . . Densher saw himself for the moment as in his purchased stall at the play; the watchful manager was in the depths of a box and the poor actress in the glare of the footlights. . . . He struck himself as having lost, for the minute, his presence of mind—so that in any case he only stared in silence at the older woman's technical challenge and at the younger one's disciplined face. It was as if the drama— it thus came to him, for the fact of a drama there was no blinking— was between *them*, them quite preponderantly; with Merton Densher relegated to mere spectatorship, a paying place in front, and one of the most expensive. . . . [T]he disciplined face did offer him over the footlights, as he believed, the small gleam, fine faint but exquisite, of a special intelligence. So might a practised performer, even when raked by double-barrelled glasses, seem to be all in her part and yet convey a sign to the person in the house she loved best. (pp. 271–72)

If it is exciting to witness Kate in this constrained, duplicitous, and visually "raked" position, however, it is equally a part of Merton's experience of manhood to be placed in the other position, in the attendance on Milly in Venice that Kate, in sight of the other women, enforces on him.

> What he could as little contrive to forget was that he had, before the two others, as it struck him . . . done exactly as he was bidden; gathered himself up without a protest and retraced his way to the palace. Present with him still was the question of whether he looked a fool for it, of whether the awkwardness he felt as the gondola rocked with the business of his leaving it . . . had furnished his friends with such entertainment as was to cause them, behind his back, to exchange intelligent smiles. (p. 360)[13]

In his formation by Kate and Milly it is Densher, here, on whom the humiliation of an involuntary two-sidedness has devolved.

A similar point could be made, I think, about the curious image of the "spring" that recurs so regularly between Kate and Milly and more generally around Kate (the same spring presumably with which Lionel is said to have tampered). The spring (a wellspring, a panther's spring) is associated with Kate's dangerousness and her coiled-up store of reserve energy, but

---

13. The immediate occasion of Densher's attending Milly at the palace here is a request from Maud Lowder, not from Kate; but of course the policy being pursued is Kate's policy, and it is to her that he feels himself bound.

at the same time (as the spring of a trap or switchblade) with her sometimes mechanicalness, with her way of being "sharp" and "ironic." (I think,
also, we are meant to hear the metallic of "iron" in Kate's "irony.") When
Kate takes on her uncanny double appearance to Milly, it comes "by some
spring of its own" (p. 199), or again "sprang, with a perversity all its own,"
from Kate's self-containment (p. 204). Milly's secrecy with Kate, similarly,
is described as "a principle of pride relatively bold and hard, a principle
that played up like a fine steel spring at the lightest pressure of too near
a footfall" (p. 339).[14] James in the Preface describes Kate's relation with
Densher as a "trap" laid for Milly which closes round her "as the result of
her mere lifting of a latch" (p. 48). It seems worth hypothesizing that the
danger the book treats as implicit in Kate's spring, and in the spring that
uncoils between her and Milly, is the danger of a clitoral eroticism (read as
mechanical, fatal, like the loaded gun in Emily Dickinson) that can be read
as conferring an ironic, enabling, by some accounts chilling distance from
the heterosexual imperative. Both Kate's autoerotic and her homoerotic
valences would be strengthened by a faculty (a faculty represented by cold
steel) that resides distinctly to one side of the novel's ostensible phallocen-
tric telos. Such a siting of her pleasure might explain something about her
way of seeming at the same time to embody "sexuality" itself for the novel
and yet to remain, in its teleologic presentation, "frigid."

And again, I would emphasize that it is in relation to the "sharp" "irony"
of Kate's homoerotic intentness that Densher's masculinity and heterosexuality, his self-perception as "straight," define themselves. As he reflects
in Venice about his position between Kate and Milly,

> It was really a matter of nerves; it was exactly because he was nervous
> that he *could* go straight; yet if that condition should increase he must
> surely go wild. He was walking in short on a high ridge, steep down on
> either side, where the proprieties—once he could face at all remain-
> ing there—reduced themselves to his keeping his head. It was Kate
> who had so perched him, and there came up for him at moments, as
> he found himself planting one foot exactly before another, a sensible
> sharpness of irony as to her management of him. . . . There glowed
> for him in fact a kind of rage at what he wasn't having; an exaspera-
> tion, a resentment, begotten truly by the very impatience of desire, in
> respect to his postponed and relegated, his so extremely manipulated
> state. (p. 362)

14. The Penguin edition says "football," but that has to be a misprint.

The "impatience of desire" sharpened and pointed by this *ressentiment* is what comes to constitute Densher's heterosexuality. "Our marriage," he insists to Kate, "will—fundamentally, somehow, don't you see?—right everything that's wrong, and I can't express to you my impatience. . . . We shall be so right" (p. 472). But the condition of that marriage, the "test" (p. 505) that he imposes on Kate to determine her fitness for it, is the exclusion from it of Milly, in the form of Milly's money. Even as the marriage falls through, the novel in its identification with Densher moves decisively toward dyadic consolidation. What its ending enforces is the end of the circle of petticoats, the strict exclusion of Kate's bond with Milly (figured as money) from Densher's bond with Milly (figured as memory). Once again, sex and gender in this novel take their shape—even when it is the most orthodox and repudiatory of shapes—from attachments of desire, identification, and disavowal involving not the regulation couple of "father" and "mother," but rather tutelary figures engaged in nonheterosexual and nondyadic relations.

In making this argument, I have so far been participating in psychoanalytic narrative structures in the sense of presenting my description of *Wings of the Dove* as a series of genetic accounts of sexing and gendering—of where gender and sexuality come from. I have focused on queer "parents" rather than queer "children" because I see an urgency in understanding queer people as not only *what the world makes* but *what makes the world;* it is time for genetic narratives like psychoanalysis (and, of course, all narratives are genetic narratives) to stop representing the idiot perseveration of the assaultive and sinister question, Where do homosexuals come from. The complications I have been trying to introduce here are a way of saying, "Get used to it." But, of course, the intense intellectual appeal of psychoanalysis is that as the modular science of the phallus, complete with a transformational grammar for translating every organ, every behavior, every role and desire into a calculus of phallic presence or absence, it not only facilitates but virtually compels an irreducible theoretical elegance in its argumentative structure. I feel painfully, on the other hand, how much economy of argument is lost in the attempt to undo the mutual transparency from organ through identity—by disarticulating from one another the histories of identity, the magnetic contingencies of organ.

A final example of this resistant disarticulation: "We shall be so right," Merton Densher fantasizes about his marriage to Kate Croy, the marriage that is to "right everything that's wrong" in certifying the young man as "straight." Why does a Bob Dylan lyric spring to mind at this point—

the funny and haunting song "You're Gonna Make Me Lonesome When You Go," with its goodbye-yellow-brick-road theme of repudiating multiple and deviant loves ("mine've been like Verlaine's with Rimbaud") and committing himself to the one true, evidently female love?

> Dragon clouds so high above,
> I've only known careless love,
> It always has hit me from below.
> But this time round it's more correct,
> Right on target, so direct—
> You're gonna make me lonesome when you go.

Part of the formal interest of this lyric lies in the rhyme that isn't a rhyme, "correct" with "direct." The hammering double iteration of the same syllable, /rect/, the syllable that (from the Latin *rectus*) signifies "straight," seems to enforce as if conclusively the song's forward trajectory toward normative (hetero)sexuality. At the same time, though, the way the lyric not only calls attention to the syllable /rect/ but makes a little drama out of tripping over it (what if the hammer's a stammer?) takes the risk of evoking that other derivative of the Latin *rectus*, the rectum—diverting the song's trajectory back to a brooding on the now disavowed form of love, figured as anal, that used to "hit me from below."

There is another similar verbal effect in *The Wings of the Dove* immediately after Kate has visited Densher in his rooms in Venice for their one sexual encounter. Densher is exultant, not to say incredulous: "It had simply *worked*, his idea, the idea he had made her accept; and all erect before him, really covering the ground as far as he could see, was the fact of the gained success that this represented. It was, otherwise, but the fact of the idea as directly applied, as converted from a luminous conception into an historic truth" (p. 399). One could scarcely imagine a more triumphally phallic image of Densher's carrying all, as who should say, before him toward a finally consolidated straightness—not just in the echoic pairing erect/direct, but even in his masterful obstetrical ability to deliver as "historic truth" what had begun as only "a luminous conception."

Yet *The Wings of the Dove* is so steeped in the thematics and aesthetics of anality that even in this momentary apotheosis of (what Kaja Silverman refers to as) "phallic rectitude" (p. 157), a lingering allusion to the pleasures of the anus seems to adhere not only to the stammering repeat syllable in erect/direct, but even to the realm of the obstetric. If there is one word that occurs even more often in the novel than the ubiquitous "queer," it is, perhaps surprisingly, "straight"—at the quickest glance I can find almost a hundred usages. "Perhaps surprisingly" unless the novel creates a junction at which straight *means* queer, where the primacy of the rectum makes it available as a switch signal for virtually all the novel's powerful sex and gender meanings. In moving toward the conclusion of this essay, I want to look briefly at some of the ways that specifically anal eroticism functions as the locus of queer tutelage in *The Wings of the Dove*—and at some of the ways, again, in which the presuppositions of psychoanalysis actually seem to traduce the multiple possibilities opened up by its interpretive tools.

The path that Silverman's essay takes toward a discussion of Jamesian anal eroticism begins with a discussion of his much-proclaimed penchant for (in his own persistent phrase) "going behind" his characters, for routing the authorial point of view austerely through the eyes of characters as they in turn view other characters—so that, Silverman suggests, a kind of daisy chain of rear penetrations may be constituted, a chain along which may run identifications and desires among reader, author, and characters. The origin of this "phantasmatic," again, Silverman traces to a primal scene in which the male child sees his "mother" and his "father" engaged in what is, or is interpreted as, anal sex. Silverman follows Freud in describing the anus as "an erotogenic zone which is undecidable with respect to gender" (p. 158), and which therefore offers the potential for sexual combinations that might not be rigidly schematized along the axis of same/different, or even of phallic/castrated. Her account offers, as we have seen, two intersecting paths over which the "child's" (James's) desires emerge as homoerotic: through a "negative Oedipal" identification with 'the mother' involving "the desire to be used by the 'father' as [the 'mother'] is"; and through "a 'pederastic identification' with the 'father,' an identification which permits the phantasizing subject to look through that figure's eyes and to participate in his sexuality by going 'behind' [i.e. genitally penetrating] him" (pp. 164–65). It is the fact that the anus is not the property of a single gender, then, that indeed makes for the possibility of homosexual emergence in Silverman's account. But is the rectum nonetheless straight? The answer, in this psychoanalytic framework, would have to be yes. Homosexuality is thinkable by this account, and thinkable even—this is the originality of Silverman's contribution—apart from the heterosexualizing presumption of gender "inversion." The problem is that it is still not thinkable if it involves only people of the same gender. In neither of these homosexual scenarios, Silverman says, "is the 'mother' dispensable. In the first instance she designates the point of interpellation, as Althusser would say, and in the second she is the necessary third term within a relay . . . which in fact begins with the 'father's' penetration of the 'mother.'" It is from this that Silverman concludes that the Jamesian phantasmatic "enclose[s] homosexuality within heterosexuality, and heterosexuality within homosexuality" (p. 165). If the mother is not dispensable, moreover, how much less so is the father. Anal sexuality cannot happen in the absence of a figure labelled "female" and a fortiori, it would seem, cannot happen in the absence of a figure labeled "male": anal sex between men is never really between men, and anal sex between women is simply beyond imagination. It seems telling to me that the most common among Silverman's somewhat weird repertoire of ways to refer to anal sex is by variations on the dated locution that refers to "using [a man] as a woman," or "using [a woman] as a man." Interestingly, to "use a man as a woman" means *exactly the same thing* as to "use a woman as a man"—i.e., to penetrate them anally. The point of this strange and insisted-upon euphemism can only be in the way it signposts with gender and with heterosexuality (on however nonsensical a pretext) the entrance to an organ and to forms of pleasure that need not have very much to do with either.

The Wings of the Dove can, I think, help us offer some resistances to the rectification of the rectum implicit in Silverman's project. To begin with, we should say that the supposed indispensability of "the mother," of the female term, in that founding phantasmatic of the daisy chain is an absolute tautology. It is nothing more than an artifact of locating this phantasmatic in the psychoanalytic framework of "the primal scene" in the first place—a scene which must occur, by definition, between "the mother" and "the father." There is no element in the scene as Silverman describes it that would be any different if the sexual act as originally viewed were to occur between two men. The supposed "indispensability" of "the mother" in psychoanalytic models of male homosexuality is the long shadow cast by the heterosexist presumption that only erotic *outcomes* may be homosexual—that erotic *origins* must uniformly be traceable to the procreative heterosexual dyad. As for the indispensability of "the father" to the scene of anal penetration, and the consequent unthinkability that the content of that scene might be lesbian, it follows of course from the same heterosexist Oedipal presumption, but follows also and even more rigorously (or tautologically) from the unswervingly phallic way that sexual acts are envisioned throughout psychoanalysis, including in Silverman's scenario. Notoriously, if sex means penetration, and penetration means penis, then there's no sex in the absence of a penis or penis prosthesis, and no sex between or among women except insofar as we may simulate having the phallus.

Now I'm not going to get started on the phallus. Personally, I can take it or leave it. But one misses a lot of the action and possibility—including lesbian action and possibility—in James's anal poetics if one is too facile

in translating every image of penetration into an image of phallic penetration. As it happens James's own, extremely dense, and highly charged
associations concerning the anus did not cluster around images of the phallus. They clustered around the hand.

In a footnote to a previous essay on James, "The Beast in the Closet,"
I quoted a passage from James's Notebooks written during a 1905 visit
to California, which still seems to me the best condensation of what *The
Wings of the Dove* presses us to recognize as his most characteristic and
fecund relation to his own anal eroticism.

> I sit here, after long weeks, at any rate, in front of my arrears, with an
> inward accumulation of material of which I feel the wealth, and as to
> which I can only invoke my familiar demon of patience, who always
> comes, doesn't he?, when I call. He is here with me in front of this
> cool green Pacific—he sits close and I feel his soft breath, which cools
> and steadies and inspires, on my cheek. Everything sinks in: nothing
> is lost; everything abides and fertilizes and renews its golden prom-
> ise, making me think with closed eyes of deep and grateful longing
> when, in the full summer days of L[amb] H[ouse], my long dusty ad-
> venture over, I shall be able to [plunge] my hand, my arm, *in,* deep and
> far, up to the shoulder—into the heavy bag of remembrance—of sug-
> gestion—of imagination—of art—and fish out every little figure and
> felicity, every little fact and fancy that can be to my purpose. These
> things are all packed away, now, thicker than I can penetrate, deeper
> than I can fathom, and there let them rest for the present, in their
> sacred cool darkness, till I shall let in upon them the mild still light of
> dear old L[amb] H[ouse]—in which they will begin to gleam and glit-
> ter and take form like the gold and jewels of a mine.[15]

At the time, I quoted this as a description of "fisting-as-*écriture*"; I am
sure it is that, but the context of *The Wings of the Dove* brings out two
other saliences of this scene of fisting equally strongly—saliences related
to each other and, of course, also to the writing process. These involve,
first, wealth, and second, parturition. One of the most audible intertexts in
the passage is surely "Full fathom five thy father lies"—with the emphasis, perhaps, on "five," the five of fingers.[16] The other important intertext

15. Henry James, *Notebooks of Henry James*, ed. F. O. Matthiessen and Kenneth B. Murdock
(New York: Oxford University Press, 1947), p. 318. Quoted in *Epistemology of the Closet*,
p. 208.
16. When the phrase "rich and strange" occurs explicitly in *Wings of the Dove*, for instance
—at the dinner party of Milly's first great success—what she is noticing as she "thrill[s],

seems to be from Book IV of the *Dunciad,* the passage where Annius describes the Greek coins he has swallowed to protect them from robbers, and anticipates their being delivered, in the course of nature, from "the living shrine" of his gut to the man who has bought them from him:

> this our paunch before
> Still bears them, faithful; and that thus I eat,
> Is to refund the Medals with the meat.
> To prove me, Goddess! clear of all design,
> Bid me with Pollio sup, as well as dine:
> There all the Learn'd shall at the labour stand,
> And Douglas lend his soft, obstetric hand.[17]

In the context of *The Dunciad,* the obstetric hand feeling for wealth in the rectum seems meant to represent the ultimate in abjection and gross-out, but under the pressure of James's brooding it has clearly undergone a sea change to become a virtually absolute symbol of imaginative value. (As such, it obviously more than complicates any reading of the novel based on a moralistic deprecation of the mercenary motive.) It is evoked in all the novel's tutorial relations, from the most demeaned to the most dignified. Of course, its easiest referent is Milly, with her combination of apparently unfathomable wealth and an unspecified illness that can at certain moments be figured as if it were a pregnancy.[18] The clearest evocation of the whole image-cluster is probably the one attached to Eugenio, "for

---

"flushe[s]," "turn[s] pale again" is the "sharp"ness of the "ring" of the air; the circumference of her attention proceeds as if hallucinatorily outward from "the smallest of things, the faces, the hands, the jewels of the women" (p. 149).

17. *The Poems of Alexander Pope,* ed. John Butt (New Haven, Conn.: Yale University Press, 1963), p. 787 (Book IV, ll. 387–94).

18. Consider this language, for instance, from the Preface: "Yes then, the *case prescribed* for its central figure a sick young woman, at the whole course of whose disintegration and the whole ordeal of whose consciousness one would have quite honestly to *assist.* The expression of her state and that of one's intimate relation to it might therefore well need to be discreet and ingenious. . . . Why had one to look so straight in the face and so closely to cross-question that idea of making one's protagonist '*sick*'?—as if to be menaced with death or danger hadn't been from time immemorial, for heroine or hero, the very shortest of all cuts to *the interesting state.* Why should a figure be disqualified for a central position by the particular circumstance that might most *quicken,* that might *crown* with a fine intensity, its liability to many accidents . . . ? *This circumstance, true enough, might disqualify it for many activities*—even though we should have imputed to it the unsurpassable activity of passionate, of inspired *resistance.* This last fact was the real *issue,* for *the way grew straight* from the moment one recognised that the poet essentially *can't* be concerned with the act of dying" (p. 36; emphasis added, with the exception of the last one).

ever carrying one well-kept Italian hand to his heart and plunging the other straight into [Milly's] pocket, which, as she had instantly observed him to recognise, fitted it like a glove" (p. 335)—the apparent (but only apparent) oddity being that Milly, while enduringly suspicious of the exquisite, responsive fit between Eugenio's hand and her pocket, also loves it and unresistingly surrenders herself to Eugenio's hands for formation. To Susan Stringham, Milly represents "a mine of something precious" ("she wasn't thinking, either, of Milly's gold") that "but needed working and would certainly yield a treasure" (p. 136). Milly's most sacred initiations occur at the hands of Sir Luke Strett,[19] in the "brown old temple of truth" (p. 209) that is his office, "the commodious 'handsome' room, far back in the fine old house . . . the rich dusk of a London 'back'" (pp. 206–7). And James likes to describe himself in what might be called the puppeteer's (or physician's) position relative to Milly's "case," as well:

> I scarce remember perhaps a case—I like even with this public grossness to insist on it—in which the curiosity of "beginning far back," as far back as possible, and even of going, to the same tune, far "behind," that is behind the face of the subject [Milly], was to assert itself with less scruple. The free hand, in this connexion, was above all agreeable. . . . (p. 41)

Again, James in the Preface puts himself in Sir Luke's diagnostically probing relation to the novel as a whole (at least "the latter half, that is the false and deformed half" of it), maintaining his "free hand" for "the preliminary cunning quest for the spot where deformity has begun" (p. 47).

Part of what is so productive about the fisting image as a sexual phantasmatic is that it can offer a switchpoint not only between homo- and heteroeroticism, but between allo- and autoeroticism (after all, James in the Notebooks passage is imagining fisting *himself*) and between the polarities that a phallic economy defines as active and passive. In the Preface, James describes Milly's general impact on everyone around her as "the strong narrowing eddies, the immense force of suction, the general engulfment" (p. 39)—about as passive as an industrial-strength vacuum cleaner. And if Kate's constantly invoked "high" "*hand*someness" (you don't lose points for registering that subliminally as high-*handed*ness) sometimes seems consistent with a familiarly phallic, subject-object understanding of her manipulations of Milly, that doesn't explain the striking phenomenon

19. The name seems to condense "straight" (straight apparently in the sense of queer) with "obstetric"—though the person he most reminds Susan Stringham of has the resonant name, Dr. Buttrick.

of Milly's own hands, or her profoundly queer response to the Bronzino portrait whose supposed resemblance to her makes, for her, a founding climax of the novel: " 'Of course her complexion's green,' she laughed; 'but mine's several shades greener. . . . Her hands are large,' Milly went on, 'but mine are larger. Mine are huge' " (p. 197). Consider, as well, how active and passive are confounded in the Preface to The Golden Bowel, I mean *The Golden Bowl,* when James describes his beloved practice of revision, "the finer appeal of accumulated 'good stuff' and . . . the interest of taking it in hand at all": "my hands," he says, "were to feel themselves full; so much more did it become a question, on the part of the accumulated good stuff, of seeming insistently to give and give." [20]

I also think there is an argument to be made that James's anal erotics function especially saliently at the level of sentence structure. I'd specifically want to point to two kinds of sentence: what I think of as the typifying sentence of *The Ambassadors,* whose relatively conventional subject-verb-object armature is disrupted, if never *quite* ruptured, as the sac of the sentence gets distended by the insinuation of one more, and just one more, and another, another, and impossibly just one *more,* qualifying phrase or clause; and the sentence of *The Wings of the Dove,* which, placing the reader less in identification with the crammed rectum and more in identification with the probing digit, presents to the reader of the beginning of the sentence a blankly baffling, "closed" grammatical facade, which yet as one arduously rounds a turn of the sentence will suddenly open out into a clear, unobstructed, and iron-strong grammatical pathway of meaning.

The obstetric hand, the fisted bowel in *The Wings of the Dove* materialize as if holographically in the convergence of two spatialities: the spatiality (which may be distinctively related to Milly) of insides and outsides, and the spatiality (which we have seen to be propagated most by Lionel and Kate Croy) of "aspect," of presented and averted, of front and back.[21] They

---

20. Henry James, *The Golden Bowl* (Harmondsworth, Middlesex: Penguin Books, 1973), p. 19.

21. The two spatialities come together on the very first page of the novel, where Kate sees herself, or invents herself, averted and presented, as the opposite of "the vulgar little street" on which her father lives, a street whose "main office was to suggest to her that the narrow black house-fronts, adjusted to a standard that would have been low even for backs, constituted quite the publicity implied by such privacies. One felt them in the room exactly as one felt the room . . . in the street" (p. 55). "To feel the street, to feel the room"—"this whole vision was the worst thing yet" (p. 55): precisely that "worst" from which the divided beauty she turns to scrutinize in the mirror is said to be "her nearest approach to an escape." The squalor of Lionel Croy's scandalous fate, in which private and public have become indistin-

go together like recto and rectum. In the Preface, James suggests that in the novel's structure Milly's plot and Kate's plot are themselves two sides of an engraved and fingered coin:

> [C]ould I but make my medal hang free, its obverse and its reverse, its face and its back, would beautifully become optional for the spectator. I somehow wanted them correspondingly embossed, wanted them inscribed and figured with an equal salience; yet it was none the less visibly my "key," as I have said, that though my regenerate young New Yorker [Milly], and what might depend on her, should form my centre, my circumference was every whit as treatable. . . . Preparatively and, as it were, yearningly—given the whole ground—one began, in the event, with the outer ring, approaching the centre thus by narrowing circumvallations. . . . (p. 40)

To make any sense of how a geography of the concentric, involving a "key" and the penetration of rings inner and outer, supervenes in this passage on a flat, two-sided geography of obverse and reverse, virtually requires that obverse and reverse be read as recto and verso—and that "recto" as the frontal face be understood as opening freely onto "rectum" as the penetrable rear. If indeed "face" and "back" "beautifully become optional for the spectator," that is because recto and verso, the straight and the "turned" or perverted, converge so narrowly onto what is not a mere punning syllable, but rather an anatomical double entendre whose interest and desirability James (and I can only join him in this) appears to have experienced as inexhaustible.

---

guishable, seems to be figured here in the two conflated spatialities: the contaminating way in which fronts have gotten mixed up with backs, in the design of these houses; the contaminating fact that insides can be felt from outside and that the outside, the street and the world, is equally palpable from the inside, the room. But notice how similar they are, finally, to the cherished and value-conferring conditions we have already seen described in Sir Luke's "brown old temple of truth" in "the rich dusk of a London 'back,'" (p. 209).

"Is the Rectum Straight?" was written in 1991. Its title is a little *hommage* to Leo Bersani. My thinking about *The Wings of the Dove* owes a great deal to Michael Moon, together with whom (and at whose instigation) I taught the novel in a graduate course at Duke during that year—and much, as well, to the people who participated in that class, including Jonathan Goldberg.

MEMORIAL

FOR

CRAIG

OWENS

Craig Owens died of AIDS-related illness in Chicago on 4 July, 1990, at the age of thirty-nine. A memorial was held later that month at The Artists' Space in New York.

Craig's and my relation was fragmentary and public. Neither of us had a right to be surprised if this fairly strange—*not* to say rare—form of love, the love of part-objects, snatches of print, glimpses and touches of a largely unfamiliar body, offered each of us a durable motive that went very deep. Craig entered my life in the most seductive of guises: in print, and as someone who "understood me." He did me the incredible honor of finding my work usable at the experimental intersection of feminist and gay identities and politics where I had so much, but at that time so fearfully and ignorantly, wanted it to tell. I *think* it was very characteristic of Craig—at least it was of our relations—that he didn't send this essay of his to me; I found it in a book in a bookstore; it was almost an afterthought when we managed to meet a while later, in the fall of '87. But then seeing him felt like keeping an assignation we'd always had, and it felt that way to me each of the few times we got to see each other. What marked Craig's presence for me was most of all the repeated proffer of amazingly intimate comfort—his proffer of it to me, I mean—in the context of really not all that much intimacy. All the different surfaces that make a self for most of us, printed pages, "our" ideas, institutional relations and activism, vibrations of a voice, the gaping abstractions and distractions of creativity, the weird holographic projections of our names and public personae, the visible and impressible extent of the parts of our bodies: it was as if these were so many ragged leaves that could be shuffled and scattered freely between me and him—all interruption, even all resistance—that still when lit up by the arc of Craig's vitality could make relation. I wonder if a lot of people didn't have similar relations with Craig—not the people he was most intimate with, but others of us for whom he animated with both his presences and absences this strange, utterly discontinuous, projective space of desire euphemistically named friendship, love at a distance, or even just reading and writing.

One of the luxuries this relation didn't have room for was the knitting together of that elastic tissue of a differential social identity that would have let me think things like, This was characteristic of Craig, this was just like Craig. I'm starving to hear that language tonight, because I don't know what Craig was "like" in those senses. Most of the things that had a chance to happen at all between us happened just once, before the moment last year when, I guess, the body constrained the soul once and for all to select her own society, then shut the door. I'm sure I'm not the only person who didn't know, when the door shut, what parts of me might be inside it, inside Craig. I could just hope there were some and that he found them to be stimulating or loving ones. Similarly I wasn't sure what it meant that so many shards of Craig's nerve and brilliance, and even body, the hilarious trenchancy of that prehensile upper lip, the comfort to be found in his telescopic arms, had become so embedded in my own identificatory life.

We all know how the climate of grief and mourning is made up of microecologies. Darkness that thickens and thickens over one of us may at the same moment be scudding apart in turbulent shreds around another. Right now I am feeling trapped in a disbelief about Craig's death that just surprises me. Part of it is the nauseatingly familiar blankness: that someone whom so many of us saw as so self-evidently treasurable, could be in a society that so failed to treasure him; a society that hated and hates so much—beginning with the great gift of a sexuality—that was most himself; a society that finally found it (to put it no more strongly than this) so possible, so little painful to let Craig die. Partly and more locally it's that I can't imagine yet what will happen to the motive Craig provided inside me. Three weeks ago I was halfway through writing an essay whose intellectual motive came, as it happened, from a couple of cryptic paragraphs of Craig's that I had been worrying over for a long time; and whose writerly motive, when the essay turned out to be much harder to think through and to enjoy than I'd expected, came *entirely* from the fun of imagining sending it to him if I could ever finish it. I didn't even picture getting an answer from him; I just found that the only pleasure of the thing came in invoking Craig's eyes to read it through. And then suddenly I couldn't do that.

For me in this part-object relation it's as if, three years ago, some incisive projectile that is Craig had gotten permanently lodged in my heart, long before I could learn who this person, or what this relation, was or could be. The eventual unfolding of that enclosure, already internal, was part of what, I always thought, was going to constitute me: for better and maybe also worse, for comfort and also conceivably danger. The verbal aura that attached to this cherished adhesion hovered around the magical words, enigmatic, magnetic. I knew enough from the start to feel privileged and grateful that this should have happened. I was most of all deeply excited. I felt fearful or paranoid, or obscurely ashamed of *myself*, sometimes when Craig and I were out of touch, as if, say, one of my holograms might have quarreled with one of his holograms without telling me about it. But it never did occur to me that this missile of relation wouldn't just unfold and unfold in me, in its own good time, its disclosures of sharpness, fragrance, pungency and surprise. Now the sense of gratitude and luck, which painfully can't diminish, are fermenting around I can't tell what point of adhesion—since I find I genuinely don't know if this inexorable disclosure now can unfold anything but, repeatedly, the loss of its subject and object.

CROSSING

OF

DISCOURSES

JANE AUSTEN

AND THE

MASTURBATING

GIRL

The phrase itself is already evidence. Roger Kimball in his treatise on educational "corruption," *Tenured Radicals,* cites the title "Jane Austen and the Masturbating Girl" from an MLA convention program quite as if he were Perry Mason, the six words a smoking gun.[1] The warm gun that, for the journalists who have adopted the phrase as an index of depravity in academe, is happiness—offering the squibby pop (fulmination? prurience? funniness?) that lets absolutely anyone, in the righteously exciting vicinity of the masturbating girl, feel a very pundit.[2]

There seems to be something self-evident—irresistibly so, to judge from its gleeful propagation—about use of the phrase, "Jane Austen and the Masturbating Girl," as the Q.E.D. of phobic narratives about the degeneracy of academic discourse in the humanities. But what? The narrative link between masturbation itself and degeneracy, though a staple of pre-1920s medical and racial science, no longer has any respectable currency. To the contrary: modern views of masturbation tend to place it firmly in the framework of optimistic, hygienic narratives of all-too-normative individual development.

1. Kimball, *Tenured Radicals,* pp. 145–46.

2. See, for a few examples of the phrase's career in journalism, Roger Rosenblatt, "The Universities: A Bitter Attack . . . ," *New York Times Book Review,* 22 April 1990, p. 3; letters in the *Book Review,* 20 May 1990, p. 54, including one from Catharine R. Stimpson disputing the evidential status of the phrase; Richard Bernstein, "The Rising Hegemony of the Politically Correct: America's Fashionable Orthodoxy," *New York Times,* 28 October 1990, sec. 4 (Week in Review), pp. 1, 4.

When Jane E. Brody, in a recent "Personal Health" column in the *New York Times*, reassures her readers that, according to experts, it is actually entirely possible for people to be healthy *without* masturbating; "that the practice is not essential to normal development and that no one who thinks it is wrong or sinful should feel he or she must try it"; and that even "those who have not masturbated . . . can have perfectly normal sex lives as adults," the all but perfectly normal Victorianist may be forgiven for feeling just a little—out of breath.[3] In this altered context, the self-evidence of a polemical link between autoeroticism and narratives of wholesale degeneracy (or, in one journalist's historically redolent term, "idiocy")[4] draws on a very widely discredited body of psychiatric and eugenic expertise whose only direct historical continuity with late twentieth-century thought has been routed straight through the rhetoric and practice of fascism. But it does so under the more acceptable gloss of the modern trivializing, hygienic developmental discourse, according to which autoeroticism not only is funny—any sexuality of any power is likely to hover near the threshold of hilarity—but must be relegated to the inarticulable space of (a barely superseded) infantility.

"Jane Austen and the Masturbating Girl"—the essay, not the phrase—began as a contribution to a Modern Language Association session that the three of us who proposed it entitled "The Muse of Masturbation." In spite of the half-century-long normalizing rehabilitation of this common form of isometric exercise, the proposal to begin an exploration of literary aspects of autoeroticism seemed to leave many people gasping. That could hardly be because literary pleasure, critical self-scrutiny, and autoeroticism have nothing in common. What seems likelier, indeed, is that the literal-minded and censorious metaphor that labels any criticism one doesn't like, or doesn't understand, with the would-be-damning epithet "mental masturbation," actually refers to a much vaster, indeed foundational open secret about how hard it is to circumscribe the vibrations of the highly relational but, in practical terms, solitary pleasure and adventure of writing itself.

As the historicization of sexuality, following the work of Foucault, becomes increasingly involved with issues of representation, different varieties of sexual experience and identity are being discovered both to possess a diachronic history—a history of significant change—and to be entangled in particularly indicative ways with aspects of epistemology and of literary creation and reception. This is no less true of autoeroticism than of other

3. Jane E. Brody, "Personal Health," *New York Times,* 4 November 1987.
4. Rosenblatt, "The Universities," p. 3.

forms of sexuality. For example, the Aesthetic in Kant is both substantively indistinguishable from, and at the same time definitionally opposed against, autoerotic pleasure. Sensibility, too—even more tellingly for the example of Austen—named the locus of a similarly dangerous overlap. As John Mullan points out in *Sentiment and Sociability: The Language of Feeling in the Eighteenth Century,* the empathetic alloidentifications that were supposed to guarantee the sociable nature of sensibility could not finally be distinguished from an epistemological solipsism, a somatics of trembling self-absorption, and ultimately—in the durable medical code for autoeroticism and its supposed sequelae—"neurasthenia." [5] Similarly unstable dichotomies between art and masturbation have persisted, culminating in those recurrent indictments of self-reflexive art and critical theory themselves as forms of mental masturbation.

Masturbation itself, as we will see, like homosexuality and heterosexuality, is being demonstrated to have a complex history. Yet there are senses in which autoeroticism seems almost uniquely—or at least, distinctively—to challenge the historicizing impulse. It is unlike heterosexuality, whose history is difficult to construct because it masquerades so readily as History itself; it is unlike homosexuality, for centuries the *crimen nefandum* or "love that dare not speak its name," the compilation of whose history requires acculturation in a rhetoric of the most pointed preterition. Because it escapes both the narrative of reproduction and (when practiced solo) even the creation of any interpersonal trace, it seems to have an affinity with amnesia, repetition or the repetition compulsion, and ahistorical or history-rupturing rhetorics of sublimity. Neil Hertz has pointed out how much of the disciplinary discourse around masturbation has been aimed at discovering or inventing proprietary traces to attach to a practice which, itself relatively traceless, may seem distinctively to threaten the orders of propriety and property.[6] And in the context of hierarchically oppressive relations between genders and between sexualities, masturbation can seem to offer—not least as an analogy to writing—a reservoir of potentially utopian metaphors and energies for independence, self-possession, and a rapture that may owe relatively little to political or interpersonal abjection.

The three participants in "The Muse of Masturbation," like most of the other scholars I know of who think and write about masturbation, have been active in lesbian and gay as well as in feminist studies. This makes

5. John Mullan, *Sentiment and Sociability: The Language of Feeling in the Eighteenth Century* (New York: Oxford University Press, 1988), esp. pp. 201–40.
6. Neil Hertz, *The End of the Line* (New York: Columbia University Press, 1985), pp. 148–49.

sense because thinking about autoeroticism is beginning to seem a productive and necessary switch-point in thinking about the relations—historical as well as intrapsychic—between homo- and heteroeroticism: a project that has not seemed engaging or necessary to scholars who do not register the antiheterosexist pressure of gay and lesbian interrogation. Additionally, it is through gay and lesbian studies that the skills for a project of historicizing any sexuality have developed, along with a tradition of valuing nonprocreative forms of creativity and pleasure; a history of being suspicious of the tendentious functioning of open secrets; and a politically urgent tropism toward the gaily and, if necessary, the defiantly explicit.

At the same time, part of the great interest of autoeroticism for lesbian and gay thought is that it is a long-execrated form of sexuality, intimately and invaluably entangled with the physical, emotional, and intellectual adventures of many, many people, that today completely *fails* to constitute anything remotely like a minority identity. The history of masturbation phobia—the astonishing range of legitimate institutions that so recently surveilled, punished, jawboned, imprisoned, terrorized, shackled, diagnosed, purged, and physically mutilated so many people, to prevent a behavior that those same institutions now consider innocuousness itself—has complex messages for sexual activism today. It seems to provide the most compelling possible exposure of the fraudulence of the scientistic claims of any discourse, *including medicine,* to say, in relation to human behavior, what constitutes disease. "The mass of 'self-defilement' literature," as Vernon A. Rosario II rather mildly points out, can "be read as a gross travesty of public health education"[7]—and queer people have recently needed every available tool of critical leverage, including travesty, against the crushing mass of legitimated discourses showing us to be moribund, mutant, pathetic, virulent, or impossible. Even as it demonstrates the absolutely discrediting inability of the "human sciences" to offer any effectual resistance to the most grossly, punitively inflictive moralistic hijacking, however, the same history of masturbation phobia can also seem to offer the heartening spectacle of a terrible oppression based on "fear" and "ignorance" that, ultimately, withered away from sheer transparent absurdity. The danger of this view is that the encouragement it offers—an encouragement we can hardly forgo, so much need do we have of courage—depends

---

7. Vernon A. Rosario II, "The 19th-Century Medical Politics of Self-Defilement and Seminal Economy," presented at the Center for Literary and Cultural Studies, Harvard University, "Nationalisms and Sexualities" conference, June 1989, p. 18.

on an Enlightenment narrative that can only relegitimate the same institutions of knowledge by which the crime was in the first place done.

Today there is no corpus of law or of medicine about masturbation; it sways no electoral politics; institutional violence and street violence do not surround it, nor does an epistemology of accusation; people who have masturbated who may contract illnesses are treated as people who are sick with specific disease organisms, rather than as revelatory embodiments of sexual fatality. Yet when so many confident jeremiads are spontaneously launched at her explicit invocation, it seems that the power of the masturbator to guarantee a Truth from which she is herself excluded has not lessened in two centuries. To have so powerful a form of *sexuality* run so fully athwart the precious and embattled sexual *identities* whose meaning and outlines we always insist on thinking we know, is only part of the revelatory power of the Muse of masturbation.

2

Bedroom scenes are not so commonplace in Jane Austen's novels that readers get jaded with the chiaroscuro of sleep and passion, wan light, damp linen, physical abandon, naked dependency, and the imperfectly clothed body. *Sense and Sensibility* has a particularly devastating bedroom scene, which begins:

> Before the house-maid had lit their fire the next day, or the sun gained any power over a cold, gloomy morning in January, Marianne, only half-dressed, was kneeling against one of the window-seats for the sake of all the little light she could command from it, and writing as fast as a continual flow of tears would permit her. In this situation, Elinor, roused from sleep by her agitation and sobs, first perceived her; and after observing her for a few moments with silent anxiety, said, in a tone of the most considerate gentleness,
> "Marianne, may I ask?—"
> "No, Elinor," she replied, "ask nothing; you will soon know all."
> The sort of desperate calmness with which this was said, lasted no longer than while she spoke, and was immediately followed by a return of the same excessive affliction. It was some minutes before she could go on with her letter, and the frequent bursts of grief which still obliged her, at intervals, to withhold her pen, were proofs enough of

her feeling how more than probable it was that she was writing for the last time to Willoughby.[8]

We know well enough who is in this *bedroom:* two women. They are Elinor and Marianne Dashwood, they are sisters, and the passion and perturbation of their love for each other is, at the very least, the backbone of this powerful novel. But who is in this *bedroom scene?* And, to put it vulgarly, what's their scene? It is the naming of a man, the absent Willoughby, that both marks this as an unmistakably sexual scene, and by the same gesture seems to displace its "sexuality" from the depicted bedroom space of same-sex tenderness, secrecy, longing, and frustration. Is this, then, a hetero- or a homoerotic novel (or moment in a novel)? No doubt it must be said to be both, if love is vectored toward an object and Elinor's here flies toward Marianne, Marianne's in turn toward Willoughby. But what, if love is defined only by its gender of object-choice, are we to make of Marianne's terrible isolation in this scene; of her unstanchable emission, convulsive and intransitive; and of the writing activity with which it wrenchingly alternates?

Even before this, of course, the homo/hetero question is problematical for its anachronism: homosexual identities, and certainly female ones, are supposed not to have had a broad discursive circulation until later in the nineteenth century, so in what sense could heterosexual identities as against them?[9] And for that matter, if we are to trust Foucault, the conceptual amalgam represented in the very term "sexual identity," the cementing of every issue of individuality, filiation, truth, and utterance *to* some representational metonymy of the genital, was a process not supposed to have been perfected for another half-century or three-quarters of a century after Austen; so that the genital implication in either "homosexual" *or* "heterosexual," to the degree that it differs from a plot of the procreative or dy-

---

8. Jane Austen, *Sense and Sensibililty* (Harmondsworth, Middlesex: Penguin Books, 1967), p. 193. Further citations from this edition are incorporated in the text.
9. This is (in relation to women) the argument of, most influentially, Lilian Faderman in *Surpassing the Love of Men: Romantic Friendship and Love between Women from the Renaissance to the Present* (New York: William Morrow, 1981), and Smith-Rosenberg in "The Female World of Love and Ritual." A recently discovered journal, published as *I Know My Own Heart: The Diaries of Anne Lister (1791–1840),* ed. Helena Whitbread (London: Virago, 1988), suggests that revisions of this narrative may, however, be necessary. It is the diary (for 1817–23) of a young, cultured, religious, socially conservative, self-aware, landowning rural Englishwoman—an almost archetypal Jane Austen heroine—who formed her sense of self around the pursuit and enjoyment of genital contact and short- and long-term intimacies with other women of various classes.

nastic (as each woman's desire seems at least for the moment to do), may mark also the possibility of an anachronistic gap.[10]

In trying to make sense of these discursive transitions, I have most before me the model of recent work on Emily Dickinson, and in particular Paula Bennett's discussion of the relation between Dickinson's heteroerotic and her homoerotic poetics in *My Life a Loaded Gun* and *Emily Dickinson: Woman Poet*.[11] Briefly, Bennett's accomplishment is to have done justice, for the somewhat later, New England figure of Dickinson, to a complex range of intense female homosocial bonds, including genitally figured ones, in her life and writing—without denying the salience and power of the male-directed eros and expectation that also sound there; without palliating the tensions acted out between the two; and at the same time without imposing an anachronistically reified view of the feminist consistency of these tensions. For instance, the all-too-available rhetoric of the polymorphous, of a utopian bisexual erotic pluralism, has little place in Bennett's account. But neither does she romanticize the female-female bonds whose excitement, perturbation, and pain—including the pain of power struggle, of betrayal, of rejection—she shows to form so much of the primary level of Dickinson's emotional life. What her demanding account does enable her to do, however, is to offer a model for understanding the bedrock, quotidian, sometimes very sexually fraught female homosocial networks in relation to the more visible and spectacularized, more narratable, but less intimate, heterosexual plots of pre-twentieth-century Anglo-American culture.

I see this work on Dickinson as exemplary for understandings of such other, culturally central, homosocially embedded women authors as Austen and, for example, the Brontës. (Surely there are important generalizations yet to be made about the attachments of sisters, perhaps of any siblings, who live together as adults.) But as I have suggested, the first range of questions yet to be asked properly in this context concerns the emergence and cultural entailments of "sexual identity" itself during this period of the incipience of "sexual identity" in its (still incompletely interrogated) modern senses. Indeed, one of the motives for this project is to denaturalize any presumptive understanding of the relation of "hetero" to "homo" as modern sexual identities—the presumption, for instance, of their symmetry, their mutual impermeability, or even of their both functioning as "sexual identities" in the same sense; the presumption, as well, that "hetero" and

10. Foucault, *The History of Sexuality*, p. 1.

11. Paula Bennett, *My Life A Loaded Gun: Female Creativity and Feminist Poetics* (Boston: Beacon Press, 1986), pp. 13–94; *Emily Dickinson: Woman Poet* (Iowa City: University of Iowa Press, 1990).

"homo," even with the possible addition of "bi," do efficiently and additively divide up the universe of sexual orientation. It seems likely to me that in Austen's time, *as in our own,* the specification of any distinct "sexual identity" magnetized and reoriented in new ways the heterogeneous erotic and epistemological energies of everyone in its social vicinity, without at the same time either adequating or descriptively exhausting those energies.

One "sexual identity" that did exist as such in Austen's time, already bringing a specific genital practice into dense compaction with issues of consciousness, truth, pedagogy, and confession, was that of the onanist. Among the sexual dimensions overridden within the past century by the world-historical homo/hetero cleavage is the one that discriminates, in the first place, the autoerotic and the alloerotic. Its history has been illuminated by recent researches of a number of scholars.[12] According to their accounts, the European phobia over masturbation came early in the "sexualizing" process described by Foucault, beginning around 1700 with publication of *Onania,* and spreading virulently after the 1750s. Although originally applied with a relative impartiality to both sexes, antionanist discourse seems to have bifurcated in the nineteenth century, and the systems of surveillance and the rhetorics of "confession" for the two genders contributed to the emergence of disparate regulatory categories and techniques, even regulatory worlds. According to Ed Cohen, for example, anxiety about boys' masturbation motivated mechanisms of school discipline and surveillance that were to contribute so much to the late-nineteenth-century emergence of a widespread, class-inflected male homosexual identity and hence to the

12. Useful historical work touching on masturbation and masturbation phobia includes G. J. Barker-Benfield, *The Horrors of the Half-Known Life: Male Attitudes Toward Women and Sexuality in Nineteenth-Century America* (New York: Harper and Row, 1976); Ed Cohen, *Talk on the Wilde Side* (New York: Routledge, 1993); John D'Emilio and Estelle B. Freedman, *Intimate Matters: A History of Sexuality in America* (New York: Harper and Row, 1988); E. H. Hare, "Masturbatory Insanity: The History of an Idea," *Journal of the Mental Sciences* 108 (1962): 1–25; Robert H. MacDonald, "The Frightful Consequences of Onanism: Notes on the History of a Delusion," *Journal of the History of Ideas* 28 (1967): 423–31; John Money, *The Destroying Angel: Sex, Fitness and Food in the Legacy of Degeneracy Theory, Graham Crackers, Kellogg's Corn Flakes and the American Health History* (Buffalo, N.Y.: Prometheus, 1985); George L. Mosse, *Nationalism and Sexuality: Respectability and Abnormal Sexuality in Modern Europe* (New York: Fertig, 1985); Robert P. Neuman, "Masturbation, Madness, and the Modern Concept of Childhood and Adolescence," *Journal of Social History* 8 (1975): 1–22; Elaine Showalter, *The Female Malady: Women, Madness, and English Culture, 1830–1980* (New York: Pantheon Books, 1985); Smith-Rosenberg, *Disorderly Conduct;* and Jean Stengers and Anne van Neck, *Histoire d'une grande peur: la masturbation* (Brussels: Editions de l'Université de Bruxelles, 1984).

modern crisis of male homo/heterosexual definition. On the other hand, anxiety about girls' and women's masturbation contributed more to the emergence of gynecology, through an accumulated expertise in and demand for genital surgery; of such identities as that of the hysteric; and of such confession-inducing disciplinary discourses as psychoanalysis.

Far from there persisting a minority identity of "the masturbator" today, of course, autoeroticism per se in the twentieth century has been conclusively subsumed under that normalizing developmental model, differently but perhaps equally demeaning, according to which it represents a relatively innocuous way station on the road to a "full," i.e., alloerotic, adult genitality defined almost exclusively by gender of object choice. As Foucault and others have noted, a lush plurality of (proscribed and regulated) sexual identities had developed by the end of the nineteenth century: even the most canonical late-Victorian art and literature are full of sadomasochistic, pederastic and pedophilic, necrophilic, as well as autoerotic images and preoccupations; while Foucault mentions the hysterical woman and the masturbating child along with "entomologized" sexological categories such as zoophiles, zooerasts, automonosexualists, and gynecomasts, as typifying the new sexual taxonomies, the sexual *"specification of individuals,"* that he sees as inaugurating the twentieth-century regime of sexuality.[13] Although Foucault is concerned to demonstrate our own continuity with nineteenth-century sexual discourse, however (appealing to his readers as "we 'Other Victorians' "),[14] it makes a yet-to-be-explored difference that the Victorian multiplication of sexual species has today all but boiled down to a single, bare—and moreover fiercely invidious— dichotomy. Most of us now correctly understand a question about our "sexual orientation" to be a demand that we classify ourselves as a heterosexual or a homosexual, regardless of whether we may or may not individually be able or willing to perform that blank, binarized act of category assignment. We also understand that the two available categories are not symmetrically but hierarchically constituted in relation to each other. The identity of the masturbator was only one of the sexual identities subsumed, erased, or overridden in this triumph of the heterosexist homo/hetero calculus. But I want to argue here that the status of the masturbator among these many identities was uniquely formative. I would suggest that as one of the very earliest embodiments of "sexual identity" in the period of the progressive epistemological overloading of sexuality, the masturbator may

13. Foucault, *History of Sexuality,* 1: 105, 43.
14. Ibid., 1: 1.

have been at the cynosural center of a remapping of individual identity, will, attention, and privacy along modern lines that the reign of "sexuality," and its generic concomitant in the novel and in novelistic point of view, now lead us to take for granted. It is of more than chronological import if the (lost) identity of the masturbator was the proto-form of modern sexual identity itself.

Thus it seems likely that in our reimaginings of the history of sexuality "as" (we vainly imagine) "we know it," through readings of classic texts, the dropping out of sight of the autoerotic term is also part of what falsely naturalizes the heterosexist imposition of these books, disguising both the rich, conflictual erotic complication of a homoerotic matrix not yet crystallized in terms of "sexual identity" and the violence of heterosexist definition finally carved out of these plots. I am taking *Sense and Sensibility* as my example here because of its odd position, at once germinal and abjected, in the Austen canon and hence in "the history of the novel"; and because its erotic axis is most obviously the unwavering but difficult love of a woman, Elinor Dashwood, for a woman, Marianne Dashwood. I don't think we can bring this desire into clear focus until we also see how Marianne's erotic identity, in turn, is not in the first place exactly either a same-sex-loving one or a cross-sex-loving one (though she loves both women and men), but rather the one that today no longer exists *as* an identity: that of the masturbating girl.

Reading the bedroom scenes of *Sense and Sensibility*, I find I have lodged in my mind a bedroom scene from another document, a narrative structured as a case history of "Onanism and Nervous Disorders in Two Little Girls" and dated "1881":

> Sometimes [X. . .'s] face is flushed and she has a roving eye; at others she is pale and listless. Often she cannot keep still, pacing up and down the bedroom, or balancing on one foot after the other. . . . During these bouts X . . . is incapable of anything: reading, conversation, games, are equally odious. All at once her expression becomes cynical, her excitement mounts. X . . . is overcome by the desire to do it, she tries not to or someone tries to stop her. Her only dominating thought is to succeed. Her eyes dart in all directions, her lips never stop twitching, her nostrils flare! Later, she calms down and is herself again. "If only I had never been born," she says to her little sister, "we would not have been a disgrace to the family!" And Y . . . replies: "Why did you teach me all these horrors then?" Upset by the reproach, X . . .

says: "If someone would only kill me! What joy. I could die without committing suicide."[15]

If what defines "sexual identity" is the impaction of epistemological issues around the core of a particular genital possibility, then the compulsive attention paid by antionanist discourse to disorders *of* attention makes it a suitable point of inauguration for modern sexuality. Marianne Dashwood, though highly intelligent, exhibits the classic consciousness symptoms noted by Tissot in 1758, including "the impairment of memory and the senses," "inability to confine the attention," and "an air of distraction, embarrassment and stupidity."[16] A surprising amount of the narrative tension of *Sense and Sensibility* comes from the bent bow of the absentation of Marianne's attention from wherever she is. "Great," at one characteristic moment, "was the perturbation of her spirits and her impatience to be gone" (p. 174); once out on the urban scene, on the other hand, "her eyes were in constant inquiry; and in whatever shop the party were engaged, her mind was equally abstracted from every thing actually before them, from all that interested and occupied the others. Restless and dissatisfied every where . . . she received no pleasure from any thing; was only impatient to be at home again . . ." (p. 180). Yet when at home, her "agitation increased as the evening drew on. She could scarcely eat any dinner, and when they afterwards returned to the drawing room, seemed anxiously listening to the sound of every carriage" (p. 177).

Marianne incarnates physical as well as perceptual irritability, to both pleasurable and painful effect. Addicted to "rapidity" (p. 75) and "requiring at once solitude and continual change of place" (p. 193), she responds to anything more sedentary with the characteristic ejaculation: "I could hardly keep my seat" (p. 51). Sitting is the most painful and exciting thing for her. Her impatience keeps her "moving from one chair to another" (p. 266) or "[getting] up, and walk[ing] about the room" (p. 269). At the happiest moments, she frankly pursues the locomotor pleasures of her own body, "running with all possible speed down the steep side of the hill"

15. Démétrius Zambaco, "Onanism and Nervous Disorders in Two Little Girls," trans. Catherine Duncan, *Semiotext(e)* 4 ("Polysexuality") (1981): 30; further citations are incorporated in the text as DZ. The letters standing in place of the girls' names are followed by ellipses in the original; other ellipses are mine. In quoting from this piece I have silently corrected some obvious typographical errors; since this issue of *Semiotext(e)* is printed entirely in capital letters, and with commas and periods of indistinguishable shape, I have also had to make some guesses about sentence division and punctuation.

16. Quoted and discussed in Cohen, *Talk on the Wilde Side*, pp. 89–90.

(p. 74) (and spraining her ankle in a tumble), eager for "the delight of a gallop" when Willoughby offers her a horse (p. 88). To quote again from the
document dated 1881,

> In addition to the practises already cited, X . . . provoked the volup-
> tuous spasm by rubbing herself on the angles of furniture, by pressing
> her thighs together, or rocking backwards and forwards on a chair.
> Out walking she would begin to limp in an odd way as if she were lop-
> sided, or kept lifting one of her feet. At other times she took little steps,
> walked quickly, or turned abruptly left. . . . If she saw some shrub she
> straddled it and rubbed herself back and forth. . . . She pretended to fall
> or stumble over something in order to rub against it. (DZ, pp. 26–27)

Exactly the overresponsive centrality of Marianne's tender "seat" as a node
of delight, resistance, and surrender—and its crucial position, as well, between the homosocial and heterosocial avidities of the plot—is harnessed
when Elinor manipulates Marianne into rejecting Willoughby's gift of the
horse:

> Elinor thought it wisest to touch that point no more. . . . Opposition
> on so tender a subject would only attach her the more to her own opin-
> ion. But by an appeal to her affection for her mother . . . Marianne
> was shortly subdued. (p. 89)

The vision of a certain autoerotic closure, absentation, self-sufficiency in
Marianne is radiantly attractive to almost everyone, female and male, who
views her; at the same time, the same autoerotic inaccessibility is legible to
them through contemporaneous discourses as a horrifying staging of auto-
consumption. As was typical until the end of the nineteenth century, Marianne's autoeroticism is not defined in opposition to her alloerotic bonds,
whether with men or with women. Rather, it signifies an excess of sexuality altogether, an excess dangerous to others but chiefly to herself: the
chastening illness that ultimately wastes her physical substance is both the
image and the punishment of the "distracted" sexuality that, continually
"forgetting itself," threatens, in her person, to subvert the novel's boundaries between the public and the private.

More from the manuscript dated 1881:

> The 19th [September]. Third cauterisation of little Y . . . who sobs and
> vociferates.
>   In the days that followed Y . . . fought successfully against tempta-

tion. She became a child again, playing with her doll, amusing herself and laughing gayly. She begs to have her hands tied each time she is not sure of herself. . . . Often she is seen to make an effort at control. Nonetheless she does it two or three times every twenty-four hours. . . . But X . . . more and more drops all pretense of modesty. One night she succeeds in rubbing herself till the blood comes on the straps that bind her. Another time, caught in the act by the governess and unable to satisfy herself, she has one of her terrible fits of rage, during which she yells: "I want to, oh how I want to! You can't understand, Mademoiselle, how I want to do it!" Her memory begins to fail. She can no longer keep up with lessons. She has hallucinations all the time. . . .

The 23rd, she repeats: "I deserve to be burnt and I will be. I will be brave during the operation, I won't cry." From ten at night until six in the morning, she has a terrible attack, falling several times into a swoon that lasted about a quarter of an hour. At times she had visual hallucinations. At other times she became delirious, wild eyed, saying: "Turn the page, who is hitting me, etc."

The 25th I apply a hot point to X's clitoris. She submits to the operation without wincing, and for twenty-four hours after the operation she is perfectly good. But then she returns with renewed frenzy to her old habits. (DZ, pp. 33)

As undisciplined as Marianne Dashwood's "abstracted" attention is, the farouche, absent presence of this figure compellingly reorganizes the attention of others: Elinor's rapt attention to her, to begin with, but also, through Elinor's, the reader's. *Sense and Sensibility* is unusual among Austen novels not for the (fair but unrigorous) consistency with which its narrative point of view is routed through a single character, Elinor; but rather for the undeviating consistency with which Elinor's regard in turn is vectored in the direction of her beloved. Elinor's self-imposed obligation to offer social countenance to the restless, insulting, magnetic, and dangerous abstraction of her sister constitutes most of the plot of the novel.

It constitutes more than plot, in fact; it creates both the consciousness and the privacy of the novel. The projectile of surveillance, epistemological demand, and remediation that both desire and "responsibility" constrain Elinor to level at Marianne, immobilized or turned back on herself by the always-newly-summoned-up delicacy of her refusal to press Marianne toward confession, make an internal space—internal, that is, to Elinor, hence to the reader hovering somewhere behind her eyes—from which there is no escape but more silent watching. About the engagement she is said to assume to exist between Marianne and Willoughby, for example, her "wonder"

> was engrossed by the extraordinary silence of her sister and Willoughby on the subject. . . . Why they should not openly acknowledge to her mother and herself, what their constant behaviour to each other declared to have taken place, Elinor could not imagine.
>
> . . . For this strange kind of secrecy maintained by them relative to their engagement, which in fact concealed nothing at all, she could not account; and it was so wholly contradictory to their general opinions and practice, that a doubt sometimes entered her mind of their being really engaged, and this doubt was enough to prevent her making any enquiry of Marianne. (p. 100)

To Marianne, on the other hand, the question of an engagement seems simply not to have arisen.

The insulation of Marianne from Elinor's own unhappiness, when Elinor is unhappy; the buffering of Mariannne's impulsiveness, and the absorption or, where that is impossible, coverture of her terrible sufferings; the constant, reparative concealment of Marianne's elopements of attention from their present company: these activities hollow out a subjectivity for Elinor and the novel that might best be described in the 1980s jargon of codependency, were not the pathologizing stigma of that term belied by the fact that, at least as far as this novel is concerned, the codependent subjectivity simply *is subjectivity.* Even Elinor's heterosexual plot with Edward Ferrars merely divides her remedial solicitude (that distinctive amalgam of "tenderness, pity, approbation, censure, and doubt" [p. 129]) between the sister who remains her first concern, and a second sufferer from *mauvaise honte,* the telltale "embarrassment," "settled" "absence of mind" (p. 123), unsocializable shyness, "want of spirits, of openness, and of consistency," "the same fettered inclination, the same inevitable necessity of temporizing with his mother" (p. 126), and a "desponding turn of mind" (p. 128), all consequent on his own servitude to an erotic habit formed in the idleness and isolation of an improperly supervised youth.

The codependency model is the less anachronistic as Marianne's and Edward's disorders share with the pre-twentieth-century version of masturbation the property of being structured as addictions. (Here, of course, I'm inviting a meditation on the history of the term "self-abuse," which referred to masturbation from the eighteenth century until very recently—when it's come, perhaps by analogy to "child abuse," to refer to battering or mutilation of oneself. Where that older sense of "abuse" has resurfaced,

on the other hand, is in the also very recent coinage, "substance abuse.")
Back to 1881:

> The afternoon of the 14th of September X . . . is in a terribly over-
> excited state. She walks about restlessly, grinding her teeth. . . . There
> is foam on her lips, she gasps, repeating, "I don't want to, I don't want
> to, I can't stop myself, I must do it! Stop me, hold my hands, tie my
> feet!" A few moments later she falls into a state of prostration, be-
> comes sweet and gentle, begging to be given another chance. "I know
> I'm killing myself," she says. "Save me." (DZ, p. 30)

Although *the addict,* as a medicalized personal identity, was (as Virginia Berridge and Griffith Edwards demonstrate in *Opium and the People*)
another product of the latter nineteenth century, the hypostatization of the
notion of "will" that would soon give rise to the "addict" identity, and that
by the late twentieth century would leave no issue of voluntarity untinged
by the concept of addiction, is already in place in *Sense and Sensibility*.[17]
A concept of addiction involves understanding something called "the will"
as a muscle that can strengthen with exercise or atrophy with disuse; the
particular muscle on which "will" is modeled in this novel is a sphincter, which, when properly toned, defines an internal space of private identity by holding some kinds of material inside, even while guarding against
the admission of others. Marianne's unpracticed muscle lets her privacy
dribble away, giving her "neither courage to speak of, nor fortitude to conceal" (p. 333) the anguish she experiences. By contrast, in the moment of
Elinor's profoundest happiness, when Marianne is restored from a grave
illness, Elinor's well-exercised muscle guarantees that what expands with
her joy is the private space that, constituting her self, constitutes it also as
the space of narrative self-reflection (not to say hoarding):

> Elinor could not be cheerful. Her joy was of a different kind, and led
> to anything rather than to gaiety. Marianne restored to life, health,
> friends, and to her doating mother, was an idea to fill her heart with
> sensations of exquisite comfort, and expand it in fervent gratitude;—
> but it led to no outward demonstrations of joy, no words, no smiles.
> All within Elinor's breast was satisfaction, silent and strong. (p. 310)

17. Virginia Berridge and Griffith Edwards, *Opium and the People: Opiate Use in Nineteenth-
Century England,* 2d ed. (New Haven, Conn.: Yale University Press, 1987). For more on the
epistemology of addiction, codependency, and addiction attribution, see my chapter "Epidemics of the Will," below.

Such an apparently generalizable ideal of individual integrity, the unitary self-containment of the strong, silent type, can never be stable, of course. Elinor has constructed herself in this way around an original lack: the absentation of her sister, and perhaps in the first place the withholding from herself of the love of their mother, whom she then compulsively unites with Marianne, the favorite, in the love-drenched tableaux of her imagination. In the inappropriately pathologizing but descriptively acute language of self-help, Marianne's addiction has mobilized in her sister a discipline that, posed as against addiction, nonetheless also is one. Elinor's pupils, those less tractable sphincters of the soul, won't close against the hapless hemorrhaging of her visual attention flow toward Marianne; it is this, indeed, that renders her consciousness, in turn, habitable, inviting, and formative to readers as "point of view."

But that hypostatization of "will" had always anyway contained the potential for the infinite regress enacted in the uncircumscribable twentieth-century epidemic of addiction attribution: the degenerative problem of where, if not in some further compulsion, one looks for the will *to* will. As when Marianne, comparing herself with the more continent Elinor,

> felt all the force of that comparison, but not as her sister had hoped, to urge her to exertion now; she felt it with all the pain of continual self-reproach, regretted most bitterly that she had never exerted herself before; but it brought only the torture of penitence, without the hope of amendment. Her mind was so much weakened that she still fancied present exertion impossible, and therefore it only dispirited her the more. (p. 270)

In addition, the concept of addiction involves a degenerative perceptual narrative of progressively deadened receptiveness to a stimulus that therefore requires to be steadily increased—as when Marianne's and her mother's "agony of grief" over the death of the father, at first overpowering, was then "voluntarily renewed, was sought for, was created again and again" (p. 42). Paradoxically afflicted, as Marianne is, by both hyperesthesia and an emboldening and addiction-producing absent-mindedness ("an heart hardened against [her friends'] merits, and a temper irritated by their very attention" [p. 337]), the species of the masturbating girl was described by Augustus Kinsley Gardner in 1860 as one

> in whom the least impression is redoubled like that of a "tam-tam," [yet who seeks] for emotions still more violent and more varied. It is this necessity which nothing can appease, which took the Roman

women to the spectacles where men were devoured by ferocious beasts. . . . It is the emptiness of an unquiet and sombre soul seeking some activity, which clings to the slightest incident of life, to elicit from it some emotion which forever escapes; in short, it is the deception and disgust of existence.[18]

The subjectivity hollowed out by *Sense and Sensibility,* then, and made available *as* subjectivity for heterosexual expropriation, is not Marianne's but Elinor's; the novel's achievement of a modern psychological interiority fit for the heterosexual romance plot is created for Elinor through her completely one-directional visual fixation on her sister's specularized, desired, envied, and punished autoeroticism. This also offers, however, a useful model for the chains of reader relations constructed by the punishing, girl-centered moral pedagogy and erotics of Austen's novels more generally. Austen criticism is notable mostly, not just for its timidity and banality, but for its unresting exaction of the spectacle of a Girl Being Taught a Lesson—for the vengefulness it vents on the heroines whom it purports to love, and whom, perhaps, it does. Thus Tony Tanner, the ultimate normal and normalizing reader of Austen, structures sentence after sentence: "Emma . . . *has to be tutored* . . . into correct vision and responsible speech. Anne Elliot *has to move,* painfully, from an excessive prudence."[19] "Some Jane Austen heroines *have to learn* their true 'duties.' They all *have to find* their proper homes" (TT, p. 33). Catherine "quite literally is in danger of perverting reality, and one of the things she *has to learn* is to break out of quotations" (TT, pp. 44–45); she "*has to be disabused* of her naive and foolish 'Gothic' expectations" (TT, p. 48). [Elizabeth and Darcy] *have to learn to see* that their novel is more properly called . . ." (TT, p. 105). A lot of Austen criticism sounds hilariously like the leering school prospectuses or governess manifestoes brandished like so many birch rods in Victorian S-M pornography. Thus Jane Nardin:

> The discipline that helps create the moral adult need not necessarily be administered in early childhood. Frequently, as we have seen, it is not—for its absence is useful in helping to create the problems with which the novel deals. But if adequate discipline is lacking in childhood, it must be supplied later, and this happens only when the char-

18. Augustus Kinsley Gardner, "Physical Decline of American Women" (1860), quoted in Barker-Benfield, *Horrors of the Half-Known Life,* pp. 273–74.
19. Tony Tanner, *Jane Austen* (Cambridge, Mass.: Harvard University Press, 1986), p. 6; page numbers of remaining quotations cited in the text with TT; emphasis, in each case, added.

acter learns "the lessons of affliction" (*Mansfield Park*, p. 459). Only after immaturity, selfishness, and excessive self-confidence have produced error, trouble, and real suffering, can the adult begin to teach himself or herself the habits of criticism and self-control which should have been inculcated in childhood.[20]

How can it have taken this long to see that when Colonel Brandon and Marianne finally get together, their first granddaughter will be Lesbia Brandon?

Even readings of Austen that are not so frankly repressive have tended to be structured by what Foucault calls "the repressive hypothesis"—especially so, indeed, to the degree that their project is avowedly *anti*repressive. And these antirepressive readings have their own way of re-creating the spectacle of the girl being taught a lesson. Call her, in this case, "Jane Austen." The sight to be relished here is, as in psychoanalysis, the forcible exaction from her manifest text of what can only be the barest confession of a self-pleasuring sexuality, a disorder or subversion, seeping out at the edges of a policial conservatism always presumed and therefore always available for violation. That virginal figure "Jane Austen," in these narratives, is herself the punishable girl who "has to learn," "has to be tutored"— in truths with which, though derived from a reading of Austen, the figure of "Jane Austen" can no more be credited than can, for their lessons, the figures "Marianne," "Emma," or, shall we say, "Dora" or "Anna O."

It is partly to interrupt this seemingly interminable scene of punitive/ pedagogical reading, interminably structured as it is by the concept of repression, that I want to make available the sense of an alternative, passionate sexual ecology—one fully available to Austen for her exciting, productive, and deliberate use, in a way it no longer is to us.

That is to say, it is no longer available to us *as passion;* even as its cynosural figure, the masturbating girl, is no longer visible as possessing a sexual identity capable of redefining and reorganizing her surround. We inherit it only in the residual forms of perception itself, of subjectivity

---

20. She is remarkably unworried about any possible excess of severity: "In this group of characters [in *Mansfield Park*], lack of discipline has the expected effect, while excessive discipline, though it causes suffering and creates some problems for Fanny and Susan Price, does indeed make them into hard-working, extremely conscientious women. The timidity and self-doubt which characterize Fanny, and which are a response to continual censure, seem a reasonable price to pay for the strong conscience that even the unfair discipline she received has nurtured in her." Jane Nardin, "Children and Their Families," *Jane Austen: New Perspectives*, ed. Janet Todd, Women and Literature, New Series, vol. 3 (New York: Holmes and Meier, 1983), pp. 73–87; both passages quoted are from p. 83.

itself, of institution itself. The last time I taught *Sense and Sensibility*, I handed out to my graduate class copies of some pages from the 1981 "Polysexuality" issue of *Semiotext(e)*, pages that reproduce without historical annotation what appears to be a late-nineteenth-century medical case history in French, from which I have also been quoting here. I handed it out then for reasons no more transparent than those that have induced me to quote from it here—beyond the true but inadequate notation that even eight years after reading it, my memory of the piece wouldn't let up its pressure on the gaze I was capable of levelling at the Austen novel. I hadn't even the New Historicists' positivist alibi for perpetuating and disseminating the shock of the violent narratives in which they trade: "Deal," don't they seem tacitly but moralistically to enjoin, "deal with your own terror, your own arousal, your disavowals, in your own way, on your own time, in your own [thereby reconstituted as invisible] privacy; it's not our responsibility, because *these awful things are real*." Surely I did want to spread around to a group of other readers, as if that would ground or diffuse it, the inadmissibly, inabsorbably complex shock of this document. But the pretext of the real was austerely withheld by the informal, perhaps only superficially sensationalistic *Semiotext(e)* format, which refused to proffer the legitimating scholarly apparatus that would give any reader the assurance of "knowing" whether the original of this document was to be looked for in an actual nineteenth-century psychiatric archive or, alternatively and every bit as credibly, in a manuscript of pornographic fiction dating from any time—any time including the present—in the intervening century. Certainly plenty of the other pieces in that issue of *Semiotext(e)* are, whatever else they are, freshly minted and joltingly potent pornography; just as certainly, nothing in the "1881" document exceeds in any detail the known practices of late-nineteenth-century medicine. And wasn't that part of the shock?—the total plausibility either way of the same masturbatory narrative, the same pruriently cool clinical gaze at it and violating hands and instruments on it, even (one might add) further along the chain, the same assimilability of it to the pseudo-distantiating relish of sophisticated contemporary projects of critique. Toward the site of the absent, distracted, and embarrassed attention of the masturbatory subject, the directing of a less accountable flood of discursive attention has continued. What is most astonishing is its continuing entirely unabated by the dissolution of its object, the sexual identity of "the masturbator" herself.

Through the frame of 1881/1981 it becomes easier to see how most of the love story of *Sense and Sensibility,* no simple one, has been rendered all but invisible to most readers, leaving a dryly static tableau of discrete moralized portraits, poised antitheses, and exemplary, deplorable, or regrettably necessary punishments, in an ascetic heterosexualizing context.[21] This tableau is what we now know as "Jane Austen"; fossilized residue of the now subtracted autoerotic spectacle, "Jane Austen" is the name whose uncanny fit with the phrase "masturbating girl" today makes a ne plus ultra of the incongruous.

This history of impoverished "Jane Austen" readings is not the result of a failure by readers to "contextualize historically": a New Historicizing point that you can't understand *Sense and Sensibility* without entering into the alterity of a bygone masturbation phobia is hardly the one I am making. What alterity? I am more struck by how profoundly, how destructively twentieth-century readings are already shaped by the discourse of masturbation and its sequelae: *more* destructively than the novel is, even though onanism per se, and the phobia against it, are living issues in the novel as they no longer are today.

We can be the less surprised by the congruence as we see masturbation and the relations surrounding it as the proto-form of any modern "sexual identity"; thus as lending their structure to many vantages of subjectivity that have survived the definitional atrophy of the masturbator as an identity: pedagogic surveillance, as we have mentioned, homo-hetero divides, psychiatry, psychoanalysis, gynecology. The interpretive habits that make it so hard to register the erotics of *Sense and Sensibility* are deeply and familiarly encoded in the therapeutic or mock-therapeutic rhetoric of the

---

21. As Mullan's *Sentiment and Sociability* suggests—and not only through the evocation of Austen's novel in its title—the eponymous antithesis "sense" vs. "sensibility" is undone by, quite specifically, the way sensibility itself functions as a point of pivotal intersection, and potentially of mutual coverture, between alloerotic and autoerotic investments. Mullan would refer to these as "sociability" vs. "isolation," "solipsism," or "hypochondria." He ignores specifically antimasturbatory medical campaigns in his discussion of late eighteenth-century medicine, but their relevance is clear enough in, for example, the discussion he does offer of the contemporaneous medical phenomenology of menstruation. "Menstruation is represented as an irregularity which takes the guise of a regularity; it is especially likely to signify a precarious condition in the bodies of those for whom womanhood does not mean the life of the fertile, domesticated, married female. Those particularly at risk are the unmarried, the ageing, and the sexually precocious" (p. 226). "The paradox, of course, is that to concentrate upon the palpitating, sensitized body of the woman caught in the difficult area between childhood and marriage is also to concede the dangers of this condition—those dangers which feature, in another form, in writings on hysteria" (p. 228). In *Epistemology of the Closet*, especially pp. 141–81, I discuss at some length the strange historical career of the epithets "sentimentality" and "sensibility," in terms of the inflammatory and scapegoating mechanics of vicariation: of the coverture offered by these apparently static nouns to the most volatile readerly interchanges between the allo- and the auto-.

"1881" document. They involve the immobilizing framing of an isolated sexual subject (a subject, that is, whose isolation is decreed by her identification with a nameable sexual identity); and her staging as a challenge or question addressed to an audience whose erotic invisibility is guaranteed by the same definitional stroke as their entitlement to intervene on the sexuality attributed to her. That it was this particular, apparently unitary and in some ways self-contained, autoerotic sexual identity that crystallized as the prototype *of* "sexual identity" made that isolating embodiment of "the sexual" easier, and made easier as well a radical naturalization and erotic dematerialization of narrative point of view concerning it.

And the dropping out of sight in this century of the masturbatory identity has only, it seems, given more the authority of self-evidence to the scientific, therapeutic, institutional, and narrative relations originally organized around it. *Sense and Sensibility* resists such "progress" only insofar as we can succeed in making narratively palpable again, under the pressure of our own needs, the great and estranging force of the homoerotic longing magnetized in it by that radiant and inattentive presence—the female figure of the love that keeps forgetting its name.

"Jane Austen and the Masturbating Girl" was written in 1990, based on a paper delivered at the Modern Language Association in 1989. What I refer to as the "1881" document from *Semiotext(e)* (see n. 15, above) was later published, under less equivocal scholarly auspices, in Jeffrey Mousaieff Masson's collection, *A Dark Science: Women, Sexuality, and Psychiatry in the Nineteenth Century* (New York: Farrar, Straus and Giroux, 1986); Démétrius Alexandre Zambaco, "Masturbation and Psychological Problems in Two Little Girls," pp. 61–89, trans. Masson and Marianne Loring from "Onanisme avec troubles nerveux chez deux petites filles," *L'Encéphale*, vol. 2 (1882), pp. 88–95, 260–74.
The project sketched out in this chapter has evoked, not only the foreclosing and disavowing responses mentioned in its first paragraph, but help, encouragement, and fellowship as well: from Michael Moon, Paula Bennett, Vernon Rosario, Ed Cohen, Barbara Herrnstein Smith, and Jonathan Goldberg, among others.

EPIDEMICS

OF THE

WILL

Once upon a time, the story goes, back in the old country, some people sometimes took opium. For many of those people, opium use was functional as a form of control: it brought into realistic conformity with the material exactions of their lives their levels of concentration, their temporality, or their alertness to stimuli such as pain. For some it may have been a source of pleasure—if a vice, then a commonplace one. For all of them, it was a behavior among other behaviors.

Then, according to the Foucauldian narrative offered by Virginia Berridge and Griffith Edwards in their important book, *Opium and the People: Opiate Use in Nineteenth-Century England,*[1] something changed. Under the taxonomic pressure of the newly ramified and pervasive medical-juridical authority of the late nineteenth century, and in the context of changing class and imperial relations, what had been a question of acts crystallized into a question of identities. To paraphrase in relation to *the addict* Foucault's famous account of the invention of *the homosexual:*

As defined by [early nineteenth-century norms], [opium-eating] was a category of . . . acts; their perpetrator was nothing more than the juridical subject of them. The nineteenth-century [addict] became a personage, a past, a case history, and a childhood. . . . [His addiction] was everywhere present in him: at the root of all

1. Berridge and Edwards, *Opium and the People.*

his actions because it was their insidious and indefinitely active principle; written immodestly on his face and body because it was a secret
that always gave itself away. . . . The [opium-eater] had been a temporary aberration; the [addict] was now a species.[2]

In the taxonomic reframing of a drug user as an addict, what changes are
the most basic terms about her. From a situation of relative homeostatic
stability and control, she is propelled into a narrative of inexorable decline
and fatality, from which she cannot disimplicate herself except by leaping
into that other, even more pathos-ridden narrative called *kicking the habit*.
From being the *subject* of her own perceptual manipulations or indeed ex-
perimentations, she is installed as the proper *object* of compulsory institutional disciplines, legal and medical, that, without actually being able to
do anything to "help" her, nonetheless presume to know her better than
she can know herself—and indeed, offer everyone in her culture who is *not*
herself the opportunity of enjoying the same flattering presumption.

The assignment of a newly pathologized addict identity to users of
opium-derived substances was not, however, the end of the story. To the
gradual extension of addiction attribution to a wider variety of "drugs"
over the first two-thirds of the twentieth century there has been added the
startling coda of several recent developments: the development, in particular, that now quite explicitly brings, not only every form of substance ingestion, but more simply every form of human behavior into the orbit of
potential addiction attribution. Think of the telling slippage that begins by
assimilating food ingestion that is perceived as excessive, with alcoholism
—in the founding of, say, Overeaters Anonymous as an explicit analogue
to Alcoholics Anonymous. From the pathologizing of food consumption
to the pathologizing of food refusal (anorexia), or even of intermittent and
highly controlled food consumption (bulimia), is a short step, but a consequential one. For the demonization of "the foreign substance" that gave
an ostensible coherence to the new concept "substance abuse" is decon-
structed almost as soon as articulated: if addiction can include ingestion,
or refusal, *or* controlled intermittent ingestion of a given substance; and if
the concept of "substance" has become too elastic to draw a boundary between the exoticism of the "foreign substance" and the domesticity of, say,
"food"; then the locus of addictiveness cannot be the substance itself and
can scarcely even be the body *it*self, but must be some overarching abstraction that governs the narrative relations between them.

2. Foucault, *History of Sexuality*, vol. 1, pp. 42–43.

That abstract space where substances and behaviors become "addictive" or "not addictive": shall we call it the healthy free will? The ability to, let us say, *choose (freely) health?* We could argue, after all, that what still unites the overeater, the anorexic, and the bulimic as substance abusers is, if not anything poisonous about the substance, still the surplus of mystical properties with which the addict/abuser projectively endows it: consolation, repose, beauty, or energy that can "really" only be internal to the addicts themselves are delusively attributed to the magical supplement that can then—whether consumed or refused—operate only corrosively on the self thus self-construed as lack. In that case, *healthy free will* would belong to the person who had in mind the project of unfolding these attributes (consolation, repose, beauty, energy) out of a self, a body, already understood as containing them *in potentio*. Under this view, which indeed is by now a staple of medicalized discourse both lay and clinical, not the dieter but the exerciser would be the person who embodied the exact opposite of addiction.

But then what are we to make of the next pathologized personage to materialize out of the taxonomic frenzy of the early 1980s: the exercise addict? In the absence of any projective hypostatization of a "foreign" substance, the object of addiction here seems to be the body itself. But more accurately the object of addiction is the display of those very qualities whose *lack* is supposed to define addiction as such: bodily autonomy; self-control; will power. The object of addiction has become precisely enjoyment of "the ability to choose freely, and freely to choose health."

It seemed logically clear from the moment of this development that if exercise was addictive, nothing couldn't be; the exercise addict was really the limit case for evacuating the concept of addiction, once and for all, of any necessary specificity of substance, bodily effect, or psychological motivation. (The brain-chemical markers invoked by scientists to "explain" addiction, of course, never had more than a tautologically explanatory or diagnostic force.) And this isn't only a theoretical *aperçu* from outside the newly efflorescent community of people interested in defining addiction. To the contrary: what is startling is the rapidity with which it has now become a commonplace that, precisely, any substance, any behavior, even any affect may be pathologized as addictive. Addiction, under this definition, resides only in the *structure* of a will that is always somehow insufficiently free, a choice whose voluntarity is insufficiently pure.

Yet, like exercise, the other activities newly pathologized under the searching rays of this new addiction attribution are the very ones that late capitalism presents *as* the ultimate emblems of control, personal discretion, freedom itself: beyond the finding of a custom-made telos in work ("workaholism") there is the telos of making ostentatiously discretionary consumer choices ("shopaholism"); of enjoying sexual variety ("being sexually compulsive"); or even of being in sustained relationships ("codependency" or "relationship addiction"). As each assertion of *will* has made voluntarity itself appear problematical in a new area, the assertion of will itself has come to appear addictive. Within the last year, there has even been a spate of journalism on the theme that the self-help groups and books that have popularized this radical critique of addiction, and that promote themselves as the only way out of it, may themselves be addictive. Typically, the headlines of these articles present the suggestion as if jokingly; but the substance of most of them, to their credit, takes seriously the self-help proposition that, understood logically, the circumference of addiction attribution is nowhere to be drawn.

The self-help analysis proceeds to this conclusion with an admirable clarity and rigor that, at best, enable very trenchant descriptive and analytic work to proceed against the grain of many contemporary ideologies. For example, while by no means all of the codependency literature is feminist, and there are limits (individualism, ahistoricity) to the political leverage of even the most ambitious, still the codependency paradigm represents an incisive deconstructive tool rendered remarkably accessible to "commonsense" experiential narratives: as such it has, as numerous self-help authors have shown, a formidable feminist descriptive salience.

The analytic work that these addiction paradigms cannot do, however, is to question or interrupt their own implication in an apparently unidirectional historical process: the propaganda of free will. By "propaganda" I mean something grammatically specific: the imperative that the concept of free will be propagated. The demand for its propagation seems to require, however, not only that it spread and circulate but that it be continually displaced: even, that the concept of free will withdraw eternally across a network of contiguities to constitute, at the last, a horizon of absence whose pressure on what is present then approaches the absolute. If the epidemic of addiction attribution constitutes, in some ways, a crisis in this propaganda, in other ways it represents its direct and inexorable continuation.

So long as an entity known as "free will" has been hypostatized and charged with ethical value (a situation whose consolidating moment in the Reformation already revealed, at the same time, the structure of its dramatic foundational fractures and their appropriability to the complex needs of capitalism)—so long as "free will" has been hypostatized and charged with ethical value, for just so long has an equally hypostatized "compulsion" had to be available as a counterstructure always internal to it, always requiring to be ejected from it. The scouring descriptive work of addiction attribution is propelled by the same imperative: its exacerbated perceptual acuteness in detecting the compulsion behind everyday voluntary is driven, ever more blindly, by its own compulsion to isolate some new, receding but absolutized space of *pure* voluntarity. The late writing of Nietzsche would obviously be the best example of this contradiction: all that there is to learn from and to recognize in Nietzsche's rendering of human psychology in terms of an exquisite phenomenology of addiction—all tied to the bizarrely moralized imperative for the invention of a Will whose value and potency seem to become more absolute as every grounding possibility for its coming into existence breathtakingly recedes and recedes.

It is striking how deftly people have taught themselves and each other to manipulate these apparently unwieldy absolutes in the lived world of addiction and addiction attribution today. In twelve-step and twelve-step-like programs, the heroics of sobriety (and by using the word heroics I don't mean to diminish the sense of a heroism to whose value and difficulty I am strongly attuned) involves a skill in, paradoxically, the micromanagement of absolutes. Under the accumulated experiential pressure and wisdom of many people's lived addictions, in twelve-step programs the loci of absolute compulsion and of absolute voluntarity are multiplied. Sites of submission to a compulsion figured as absolute include the insistence on a pathologizing model ("alcoholism is an illness") that another kind of group might experience as disempowering or demeaning; the subscription to an antiexistential rhetoric of unchangeable identities ("there are no ex-alcoholics, only recovering alcoholics"); and the submissive recourse to a receding but structurally necessary "higher power." At the same time, sites of a voluntarity also figured as absolute are procured and multiplied by fragmentation—in rituals of taking responsibility for damages of the past; in the decentralized and highly egalitarian, if also very stylized, structure of group experience; and especially through a technique of temporal fragmentation, the highly existential "one day at a time" that dislinks every moment of choice (and of course they are infinite) from both the identity-history and the intention-futurity that might be thought to constrain it. Among these sites of subtle negotiation among reduplicated absolutes, many people seem able to create for themselves a workable way out of the deadly system of double binds where an assertion that one can act freely is always read in the damning light of the open secret that the behavior in question is utterly compelled—while one's assertion that one was, after all, compelled, shrivels in the equally stark light of the open secret that one might indeed at any given moment have chosen differently.

If will and compulsion are, necessarily, mutually internal categories; if the nineteenth-century isolation of addiction "itself" must therefore be seen as part of the same historical process as the Nietzschean hypostatization of Will "itself"; if this pressure of overlapping classifications does nothing more than chisel a historically specific point of stress into the centuries-old fault line of free will; then we must ask, what are the specifying historical coordinates of the present crisis of addiction attribution? The simplest answer, I think, to the question, "Why now?"—why the twentieth century, and most of all its final quarter, should turn out to be the site of this epidemic of addiction and addiction attribution—must lie in the peculiarly resonant relations that seem to obtain between the problematics of addiction and those of the consumer phase of international capitalism. From the Opium Wars of the mid-nineteenth century up to the current details of U.S. relations with Turkey, Colombia, Mexico, and Peru, the drama of "foreign substances" and the drama of the new imperialisms and the new nationalisms have been quite inextricable. The integrity of (new and contested) national borders, the reifications of national will and vitality, were readily organized around these narratives of introjection. From as far back as Mandeville, moreover, the opium product, the highly condensed, portable, expensive, commerce-intensive substance, par excellence the cartel-vulnerable commodity crop for export as opposed to the subsistence crop for home use, was seen as having a unique ability to pry the potentially unlimited trajectory of demand, in its users, conclusively and ever-increasingly apart from the relative homeostasis of need. With the advent of the age of advertising, then, the addictive substance was spectacularly available to *be brought back home* as a representation for emerging intuitions about commodity fetishism.

Furthermore, the paraphrase from Foucault's *History of Sexuality* with which I began this essay suggests that, because the emergence of the addict identity and the homosexual identity have so closely coincided both structurally and temporally, their historical interimplications may have been deep indeed. In *The Picture of Dorian Gray* as in, for instance, *Dr. Jekyll and Mr. Hyde,* drug addiction is both a camouflage and an expression for the dynamics of male same-sex desire and its prohibition: both books begin by looking like stories of erotic tensions between men, and end up as cautionary tales of solitary substance abusers. The two new taxonomies of the addict and the homosexual condensed many of the same issues for late-nineteenth-century culture: the old antisodomitic opposition between something called nature and that which is *contra naturam* blended with a treacherous apparent seamlessness into a new opposition between substances that are *natural* (e.g., "food") and those that are *artificial* (e.g., "drugs"); and hence into the characteristic twentieth-century way of distinguishing desires themselves between those considered natural, called "needs," and those considered artificial, called "addictions." Perhaps the reifying classification of certain particular, palpable substances as unnatural in their (artificially stimulating) relation to "natural" desire must ultimately throw into question the naturalness of any desire (Wilde: "Anything becomes a pleasure if one does it too often").[3] And of course one of the many echoes resounding around the terrible accident of HIV and the terrible nonaccident of the overdetermined ravage of AIDS is the way it seems "naturally" to ratify and associate—as *un*natural, as unsuited for survival, as the appropriate objects of neglect, specularized suffering and premature death—the notionally self-evident "risk group" categories of the gay man and the addict.[4]

Even apart from AIDS, some of the different connections in which addiction attribution today brings up the question of the "natural" are illustrated in the Opinion page from a 1987 *USA Today*, focusing on the controversies around steroid use by athletes. The very definition of addiction is up for grabs here: Ron Hale says steroids aren't addictive (and hence shouldn't be banned) because they "aren't uppers, they're not downers, they don't alter the mind, and it's not a habit that can't be broken"; William N. Taylor says they *are* and therefore *should*, because "self-users exhibit tolerance to the drugs and withdrawal phenomena so that personality changes and addiction are common."[5] In the all-American voices cited in this debate, the Natural is invoked to ward off bogeys that hover around us on several different axes: the cyborg axis that stretches between people and machines ("If athletes have to turn to steroids, they are trying to build up a machine and not a human being"); the narrative evolutionary axis that stretches between people and animals, only glancing at that intermediate space of interbreeding, the "era of monstrous athletes"; the axis that extends from the autoerotic to the alloerotic (since "muscular bodies are usually the re-

3. Oscar Wilde, *The Picture of Dorian Gray* (Harmondsworth, Middlesex: Penguin Books, 1981), p. 236.
4. Issues mentioned in the preceding two paragraphs are contextualized more fully in my *Epistemology of the Closet*, chap. 3, and the chapter below, "Nationalisms and Sexualities: As Opposed to What?"
5. All quotations are from *USA Today*, 6 January 1987, p. 10A.

Cartoon by David
Seavey. Copyright 1987,
USA Today. Reprinted
with permission.

sult of years of anabolic steroid-induced narcissism"); the puritanical axis
between a moralized "health" and the deprecated "enhancement of a person's body"; the axis between satiable and "insatiable" drives, whether
for sex or for substance ingestion; the axis between immanent "natural
ability" and the extrinsic additive ("individuals should compete with their
bodies' resources rather than relying on some unnatural chemical"); and
of course the axis between the "free" world, where steroid use in amateur
competition is absolutely prohibited, and "Russia," where, the editorial
speculates, it "might" be absolutely compulsory. Steroids, moreover, by
inducing " 'roid rage" have the dangerous property of blurring the should-
be-obvious distinction between heterosexuality and male violence against
women. Worse, even though steroids ought to be easy to situate on the axis
between male and female gender—they are, after all, "man-made versions
of the male hormone testosterone"—they actually seem to have unpredict-
ably destabilizing effects on gender attributes. If their generally virilizing
tendency leads to "women built like men and men built like the Carpathian
Mountains," if they cause "growth of facial hair, baldness, and shrunken
breasts" in women and "overly aggressive and violent behavior" in men,
they also induce in women a "nymphomania" that can't be so securely
categorized as a *masculinity* effect; and most damagingly to the gender
schema that always seems to underlie hormonal models, they cause men to
"develop[ ] female characteristics like breasts." The chilling editorial car-

toon summarizes the worst news about the foreign substance of steroids: the bulked-up body of absolute voluntarity, scared as it is scary, is haunted by a memento mori, the X-ray skeleton of its own absolute compulsion; between the two images, however, the supposedly self-evident place of the desirable, "natural" outlines of just plain boy has already corroded away beyond remembering or restoring.

It is not so hard, then, to come up with a proliferation of—not causal *explanations* for the present epidemic of addiction attribution—but at least aspects of the present historical moment that seem to engage the addiction model in politically fraught, discursively productive mutual relations of representation and misrepresentation. I'd like to end, though, by asking a much more difficult question than Why now?—namely, How could it have been otherwise? What shards of outdated cognitive resource may we still find scattered by the roadside of progress, resources by which it might be, or might once have been, possible to think voluntarity and compulsion differently—to resist simply re-propelling the propaganda of a receding Free Will? I have to say that I do not find the concept of the unconscious, whether individual or historical, to be of much help in approaching this particular question: both psychoanalytic and Marxist thought seem to have the modern heroics of voluntarity/compulsion already inscribed too deeply at the source of their narrative and analytic energies. When I look, on the other hand, at the work—from which I've learned a lot—of a Thomas Szasz, whose deconstructive analyses of concepts like addiction have been so vigorous that they might better be called debunking, I also see that his salutary leverage on the pathologizing mythologies of compulsion owes everything to a tropism toward the absolute of punishable free will that itself more than verges on the authoritarian.

I'll just suggest briefly that the best luck I've had so far in reconstructing an "otherwise" for addiction attribution has been through a tradition that is, not opposed to it or explanatory of it, but rather one step to the side of it. That is the tradition of reflecting on *habit*, a version of repeated action that moves, not toward metaphysical absolutes, but toward interrelations of the action—and the self acting—with the bodily habitus, the appareling habit, the sheltering habitation, everything that marks the traces of that habit on a world that the metaphysical absolutes would have left a vacuum. Though perpetuated and fairly intensively moralized from at least Cicero up to at least William James, with an especially acute psychologizing currency around, for instance, the eighteenth-century and Romantic origins of the English novel, the worldly concept of habit has dropped out of theorized use with the supervention in this century of addiction and the other glamorizing paradigms oriented around absolutes of compulsion/voluntarity. And indeed, I can understand the mistrust of modern versions of "habit," such as ego psychology, whose dependence on a metaphorics of *consolidation,* and whose consequent ethicization of the unitary self, seem to render it peculiarly vulnerable to unacknowledged forms of moralism. Yet the unmoralized usage of the language of habit in, for instance, Proust is as scouring as any version of contemporary addiction attribution—without at all requiring the hypostatization of a ghostly, punitive free will on the receding horizon. Proust treats habit as, in the first instance, a perceptual matter, which means that his wealth of resources for denaturalizing the polarities "active" and "passive" in perception is at work in all his discussion of habit, the human faculty that can "chang[e] the colour of the curtains, silenc[e] the clock, [bring] an expression of pity to the cruel, slanting face of the glass."[6] Habituation is "that operation which we must always start afresh, longer, more difficult than the turning inside out of an eyelid, and which consists in the imposition of our own familiar soul on the terrifying soul of our surroundings" (vol. 2, p. 791). A banal but precious opiate, habit makes us blind to—and thus enables to come into existence—our surroundings, ourselves as we appear to others, and the imprint of others in ourselves.

Habit also, however, demarcates the space of perceptual and proprioceptive reversal and revelation—revelation at which introspection itself can never arrive. As on the narrator's receipt of Albertine's letter saying she has left him for good:

> Yes, a moment ago, before Françoise came into the room, I had believed that I no longer loved Albertine, I had believed that I was leaving nothing out of account, like a rigorous analyst; I had believed that I knew the state of my own heart. . . . I had been mistaken in thinking that I could see clearly into my own heart. But this knowledge, which the shrewdest perceptions of the mind would not have given me, had now been brought to me, hard, glittering, strange, like a crystallized salt, by the abrupt reaction of pain. I was so much in the habit of having Albertine with me, and now I suddenly saw a new aspect of Habit. Hitherto I had regarded it chiefly as an annihilating force which suppresses the originality and even the awareness of one's perceptions; now I saw it as a dread deity, so riveted to one's being, its insignificant face so incrusted in one's heart, that if it detaches itself,

6. Marcel Proust, *Remembrance of Things Past,* trans. C. K. Scott Moncrieff and Terence Kilmartin, 3 vols. (New York: Vintage Books, 1982), vol. 1, p. 9. Proust citations in the text are to this edition.

if it turns away from one, this deity that one had barely distinguished inflicts on one sufferings more terrible than any other and is then as cruel as death itself. (vol. 3, p. 426)

Habits in Proust, like lies and foolish sorrows, resemble "servants, obscure and detested, against whom one struggles, beneath whose dominion one more and more completely falls, dire and dreadful servants whom it is impossible to replace and who by subterranean paths lead us towards truth and death" (vol. 3, p. 948). And yet it is also they, the habits "—even the meanest of them, such as our obscure attachments to the dimensions, to the atmosphere of a bedroom—that take fright and refuse, in acts of rebellion which we must recognise to be a secret, partial, tangible and true aspect of our resistance to death" (vol. 1, p. 722).

It is extraordinarily difficult to imagine an analytically usable language of habit, in a conceptual landscape so rubbled and defeatured by the twin hurricanes named Just Do It and Just Say No. I most feel the vertigo of this scene of vastation when I learn of Philip Morris's buying, or at least renting from the National Archives, sponsorship of the Bill of Rights to the U.S. Constitution, involving as well a series of magazine ads to promote the abstraction Freedom, both the Freedoms already articulated in that document and others, it is more than hinted, that may be yet to be specified. A recent ad, for instance, features a handsome black-and-white photograph of a smiling but dignified Barbara Jordan whom it quotes as saying, "The Bill of Rights was not ordained by nature or God. It's very human, very fragile." The ad is short on specifics, however. The rest of the text: "The Bill of Rights has been a source of comfort for me. While I was born into poverty, I knew it didn't have to be a permanent condition. I was free to do whatever I wanted to do. And the liberating force throughout my life and career has been the Bill of Rights. It's where the United States of America comes to life. Without it, this country as we know it would cease to exist."

I won't even discuss the hideousness of the "irony" of this propaganda for Freedom coming from an industry whose pet senator, Jesse Helms of North Carolina, murders constitutional freedoms as lavishly and recklessly as he allows the lives of gay men and addicts to be extinguished.

I assume that this advertising campaign represents more than an attempt to associate tobacco advertising, at present itself very embattled, more intimately with the First Amendment protections on which it battens whenever it is threatened with further regulation. Beyond the issue of Freedom to advertise, it alludes to past and future tobacco company campaigns on the theme of smokers' rights, campaigns (including publication of a glossy new magazine just for smokers) which are attempting to mobilize and crystallize smokers as a potent lobbying and advocacy group on the model of something like the National Rifle Association. Now, in terms of the addiction/free will paradigm, the bright idea of crystallizing a smoker advocacy community is obviously a volatile one indeed: if on the one hand, the tobacco companies cherish the vision of a population united to claim the right to smoking as a high exercise of individual freedom, in solidarity with their good friends at the tobacco companies to whom not only the articulation but in fact the setting of these assertive "rights" agendas is meant unquestioningly, even abjectly to be surrendered; —still on the other hand, that Philip Morris wet dream must be shadowed by a nightmare in which smokers could unite to claim rights, not as embodiments of that ultimate freedom, the freedom to smoke, but rather *as addicts,* as people who define themselves as *not* having freedom with respect to smoking. It is, of course, only by claiming the latter identity, by a willingness to stigmatize themselves further by making common cause with the most disempowered of social groups in demands for, shall we say, reliable, unpunitive access to the addictive substance; for affordable or free, high-quality, nonjudgmental health care available to all who have been, still are, or risk becoming addicts (i.e., to everyone); and for freedom from economic exploitation by traffickers in the addictive substance—it is only by making something like this claim to, or acknowledgment of, the pathologized addict identity that smokers, as a body, could, paradoxically, empower themselves in legal, economic, and ideological contestation *against* the tobacco companies, as well as in areas where their interests may coincide with the companies'. (The latter areas might include, for instance, limiting workplace prohibitions on smoking and resisting victim-blaming insurance surcharges against smokers.)

There is plenty of evidence that the thing tobacco companies most fear is official and legal recognition of the open secret that, behind a facade of voluntarity, smoking "is really" addictive. If the operation of that open secret is, at present, disempowering to smokers, so at the same time is the dynamic of the countervailing open secret: that anyone who claims to be compelled to smoke is actually, with every cigarette she lights, making "a free choice" not to do otherwise. I see these ads as most effectually performing, at some level, the warning taunt of a blackmailer, aimed at smokers: at driving in and in, against them, the ugly twisting point that in the present discursive constructions of consumer capitalism, the powers of our free will are always already vitiated by the "truth" of compulsion, while the powers attaching to an acknowledged compulsion are always already vitiated by the "truth" of our free will. No wonder that, as Proust suggests, acts of refusal and rebellion in this wasting landscape need to muster real rhetorical and political cunning to remain secret, partial, tangible, true.

"Epidemics of the Will" was written for a conference on "Epidemics" held at MIT in October 1990; I am grateful to David Halperin for the instigation to the paper, and am also indebted to Andrew Parker for staying on my case about it. I am also especially grateful for the exciting opportunity, over many years, of sharing with Joshua Wilner his meditations on this subject. His writing about substance ingestion and Romanticism includes "Autobiography and Addiction: The Case of De Quincey," *Genre* 14 (Winter 1981): 493–503; and "The Stewed Muse of Prose," *MLN* (December 1989): 1085–98.

NATIONALISMS

AND SEXUALITIES:

AS OPPOSED

TO WHAT?

In September 1988 I was involved with a conference at my home institution on Liberal Arts Education in the Late Twentieth Century. The speakers at the three-day conference mostly represented an emerging mainstream-left consensus on issues such as pluralization of canons and curricula. I participated in the conference in two ways. On the second day, Henry Louis Gates and I were on a panel about emerging canons, where I gave a talk about antihomophobic pedagogy and the difficult question of defining a gay canon. And on the last day, in the roundup discussion period that concluded the conference, I tried to articulate a serious unease that I had had with the whole proceeding. What was disturbing me then was the way the term "American" had come, unbidden and unremarked, to occupy a definitional center for almost every single one of the papers, and for the conference as a whole, in a way that no one could even seem to make visible enough to question. That the conference, whose title did not specify "America" and whose topic was by no means *necessarily* circumscribed by any boundaries of the national, had lined up so neatly in the current train of contestations about what exclusive or inclusive, white or nonwhite, gay or straight, homogeneous or heterogeneous visions are to constitute a *national* culture, a *national* identity, about where we are to look for the special American values—be they good or bad—of America and American education, seemed a striking datum to just the degree that it was taken for granted. It seemed to me that to the social archaeologist of the future, a conference like this one might figure not most saliently as an agonistic moment in the history of the liberal arts, nor as the nexus of a conflict between a political left and a political right, but as a moment of the heavy, uncontestable re-engraving of nationalism—or more accurately, of what Benedict Anderson refers to as "nation-ness"[1]—as the invisible outline whose unquestioned boundaries could only be strengthened by the apparent fierceness of the battles fought in its name and on its ground. Why, I asked at that final session, when we talk about all the very disparate things we have been talking about, do we always seem to find—do we always seem to fail to notice or query—that we are also talking about, and ratifying—by appealing back to different versions of it to ratify *us*— the primary realness of, of all imaginary things in the world, "America"?

A few days later, the conference was reported on in the *New York Times*. Given the framework of recent controversy over the liberal arts, the newspaper account was miraculously positive; it seemed clear that everything that could be done, had been done, to bathe the goings-on in North Carolina in a benign and edifying light. The headline, for instance, was, blandly, "Liberal Arts Scholars Seek to Broaden Their Field." The picture run with the article, for another instance, was a nicely emblematic one of me lecturing forcefully, with Skip Gates as an appreciative listener, our proximity buffered by the brow-furrowed head of a white man. The list of topics discussed, for a third instance, though voluminous, was kind enough to omit any mention whatever of so alarming a subject as homophobia or gay and lesbian culture, on which I was in fact lecturing when the picture was taken. Instead, a quotation, possibly accurate, from the last session of the conference was attached to my name, as a sort of clincher to prove the sincerity and unimpeachable good faith of all the participants of this conference. "Eve K. Sedgewick, an English professor at Duke, said: 'What we're really talking about is America, our vision of what the country is, what the country should be.' "[2]

I expect that the kind of process that went into disappearing the gay/lesbian content of that conference from the *New York Times* felt completely different from the kind of process that led to the 180-degree miscontextualization of the remark on America. Someone, I am sure—either the reporter or some person or persons elsewhere in the editorial lineup—someone must have given conscious thought to whether it would be possible to

1. Benedict Anderson, *Imagined Communities: Reflections on the Origin and Spread of Nationalism* (London: Verso Press, 1983), p. 13; further citations incorporated in the text.
2. Spelling unmodified. Lee A. Daniels, "Liberal Arts Scholars Seek to Broaden Their Field," *New York Times*, 21 September 1988, p. 23.

name gay issues in the article without disrupting the bland homogeneity of the legitimating aegis it aimed to throw over the work of the conference. Such person or persons decided, obviously, *no*—but at least I feel confident that they did have to decide, in a discursive context where, in the elegant formulation of Congressman Barney Frank, homosexuality has gone from seeming to public arbiters a "no way" issue, where there is only one articulated side and that side is negative, to being what he calls an "oh shit" issue, with two highly articulated opposing sides, each one of which is eager and able to make a federal case. Concerning "America," however, I am just as confident that nothing that felt in the least like a decision had to be made, by writer or editor, in order to misrepresent as a legitimating, populist appeal to national identity what was in fact an attempt to put into question the grounds of any such appeal. To the contrary: it can only have seemed (quite literally) the most natural thing in the world to anyone at the *New York Times* to assume, not only that any labor of intellectual legitimation must necessarily, sooner or later, be channeled through an appeal to the national, but conversely that any mention whatsoever of the national must necessarily, sooner or later, reveal itself as an appeal for legitimation.

To come to the end of this particular journalistic saga, the difference between the precisely contestable current status of gay definition and the completely uncontestable current status of national definition was brought, as they say, home to me with the next Sunday's *Times,* where the News of the Week section carried a second, this time virulently negative article about the same conference. Where our friends may politely, protectively— and of course, obliteratingly—omit the mention of gay studies and anti-homophobic constituencies, our enemies have no such hesitation about naming us; the "homosexual" component of the conference was salient in the attack on it, by a different calculation in the same arithmetic that had elided us from its defense. What the second article shared unquestioningly with the first, on the other hand, was the assumption that to discuss the future of the humanities is coextensive with discussing the Americanness of America: that the national question is both the grounding origin and the necessary terminus of any understanding whatsoever of consciousness or production, perhaps *especially* insofar as those are also carried out in the name of the universal. The second article ends with a telephone interview with Lynn V. Cheney, chairwoman of the National Endowment for the Humanities.

> "I have the conviction that great literature, no matter whom it is written by, speaks to transcendent values that we all share, no matter what

our time and circumstance," Mrs. Cheney said. . . . There is everything to be gained, she said, from studies by and about women, blacks, and other elements in American culture. Still, she maintained that American history and values derive primarily from the great thinkers of Europe, and not from Asia or Africa. "On the West," she argued, "the first responsibility is to ground students in the culture that gave rise to the institutions of our democracy."[3]

What we're really talking about is America, our vision of what the country is, what that country should be. Is there any way to stop? In his impressively suggestive book on this subject, *Imagined Communities: Reflections on the Origin and Spread of Nationalism,* the anthropologist Benedict Anderson points out that while "nationalism" is usually discussed as an ideology, like liberalism or fascism, against which an alternative ideology might be posed, to the contrary "it would, I think, make things easier if one treated it as if it belonged with 'kinship' and 'religion'" (p. 15)—by implication, as the name of an entire underlying dimension of modern social functioning that could then be organized in a near-infinite number of different and even contradictory ways. The work—I should more accurately say the struggle—of defamiliarizing and thereby visibilizing the nationalism that forms the overarching ideology of our age is difficult to the extent that one or another nationalism tends to become the form of last resort for *every* legitimizing political appeal—whether right or left, imperialist or anti-imperialist, religious or secular, elitist or populist, capitalist or anti-capitalist, cynical or utopian, and whether or not on behalf of an ideology that has any account whatever to offer of the status of the "nation" as such. (No analysis through Marxism, capitalism, religion, or race, for instance, offers any intelligible justification of national form, yet each of these kinds of power today takes national or nationalist shapes.)

My sense is that an underlying liberal understanding of nationalism as an ideology, as something *against which* there exist conceptual tools to fight, is currently shaping our sense of the relations between nationalism and sexuality in circumscriptive ways. It's characteristic of mainstream-left thought (when it is not in the grip of utopian misunderstandings of foreign nationalisms, whose *un*masking as "ordinary" nationalisms always performs another extension of the same cynical narrative) to associate nationalism in the first place and definitionally with a nineteenth-century

3. Richard Bernstein, "Academia's Liberals Defend Their Carnival of Canons Against Bloom's 'Killer B's,'" *New York Times,* 25 September 1988, p. 26E.

reactionary European project of bourgeois boundary consolidation: with rightist projects of racial, gender, sexual and other scapegoating around borders shaped by a quasifamilial ideal of purity that would at the same time distance and justify the inflictions of overseas empire. The topos of the creation, reification, and expulsion of the Other, and signally the Oriental- ized other, in the emergence of the modern European state, has become a central tool of liberal analysis; and it is the explanatory aegis of the Other or Othered that has, for the most part, allowed people of variant sexuali- ties, along with the non-Christian, nonwhite, and medically stigmatized people, to become visible in liberal narratives about the origins of nation- alism. Roughly, one could say that the trope of the Other is to current understandings of nationalism what the repression hypothesis was to pre- Foucauldian understandings of sexuality. The trope of the Other is in fact true to a certain irreducible core of suffering, denial, and crude reification whose historical and contemporary traces are everywhere to be seen and felt. Also like the repression hypothesis, however, the trope of the Other in relation to nationalism must almost a priori fail to do justice to the com- plex activity, creativity, and engagement of those whom it figures simply as relegated objects—their activity, creativity, and engagement with and on behalf of, among other things, that protean fabric of public discourse that does also figure their own relegation. Beyond the always accurate but spectacularly un-analytic diagnosis of, say, "internalized homophobia"— diagnosis to which, like the diagnosis of "Jewish" or "colonial" "self- hatred," the trope of Otherness necessarily has ultimate recourse—beyond the search for "internalized homophobia," what questions may we learn to ask about the extremely varied kinds of importance of national ques- tions in sexual politics—in those, for instance, to take some complicated examples, of a Wilde, a Forster, a Nietzsche, or a Proust?

"In the modern world," Benedict Anderson writes, "everyone can, should, will 'have' a nationality, as he or she 'has' a gender" (p. 14). The implication, I think, is that just as every culture has *some* mechanism— different mechanisms—to constitute what Gayle Rubin refers to as a "sex/ gender system,"[4] a way of negotiating back and forth between chromo- somal sex and social gender, so every *modern* culture and person must be seen as partaking of what we might (albeit clumsily) call a "habitation/ nation system." The "habitation/nation system" would be the set of discur-

---

4. Gayle Rubin, "The Traffic in Women: Notes on the 'Political Economy' of Sex," in Rayna R. Reiter, ed., *Toward an Anthropology of Women* (New York and London: Monthly Review Press, 1975), p. 159.

sive and institutional arrangements that mediate between the physical fact
that each person inhabits, at a given time, a particular geographical space,
and the far more abstract, sometimes even apparently unrelated organization of what has emerged since the late seventeenth century as her/his
national identity, as signalized by, for instance, citizenship.

Practically, the existence of habitation/nation systems, and their great
variety worldwide, tend to become visible most easily as "exceptional"
stresses on a system (thereby) taken as normal: through personal and
political crises concerning exile and expatriation, sanctuary, guest-workers
whose status is for some reason deroutinized, changes in the laws of aliyah,
emerging or resurgent nationalisms within previously established states,
etc. But let us, for a minute, consider some consequences of Anderson's
analogy between modern nationality and gender. To suggest that everyone
might "have" a nationality as everyone "has" a gender presupposes, what
may well be true, and may well always have been true, that everyone does
"have" a gender. But it needn't presuppose that everyone "has" a gender
as everyone else "has" a gender—that everyone "has" a gender in the same
way, or that "having" a gender is the same kind of act, process, or possession for every person or for every gender. It wouldn't be contradicted,
even, by something like Nancy Chodorow's finding that female and male
gender were by definition different kinds of things attainable by different kinds of process—normative female gender by a primary *identification
with* a mother, say, and normative male gender by a secondary *differentiation from* female gender. Under such a finding (and while I don't assume it's
either true or exhaustive, I don't think it particularly implausible), while
everyone could perhaps be said to *have* a gender, while gender would certainly constitute a first category of the "normal," and while there would
be an availability of so-called normal female gender and so-called normal
male gender, there nevertheless could not be any such thing as *the* normal
way to "have" a gender. Under the Chodorovian supposition, the definitional Other of gender—the answer to "gender as opposed to what?"—
would be different in each case: "normal" female gender would be *as opposed to* no gender; "normal" male gender *as opposed to* female gender.

Something similar may be true of "nation-ness" and nationality. Even a
simple exercise like attempting to compare the "nationality" of a Gypsy in
Germany (formerly East or West Germany), that of a Turkish guest-worker
in France, and that of a person who works in Johannesburg, lives in Soweto,
and has an official assignment to a "homeland" s/he may never have seen
suggests how ragged and unrationalizable may be not only the rights and

entailments, but the definitional relations of the habitation/nation systems of the modern world.

Benedict Anderson suggests that if any one thing defines the modern "nation" as opposed to "the older imagining, where states were defined by centres, borders were porous and indistinct, and sovereignties faded imperceptibly into one another," it is that "in the modern conception, state sovereignty is fully, flatly, and evenly operative over each square centimetre of a legally demarcated community" (p. 26)—and so is "that remarkable confidence of community in anonymity which is the hallmark of modern nations" (p. 40). Thus it should not surprise you when at the Buffalo airport, waiting for your little commuter flight to Toronto, and starting to wonder what the weather there is likely to be like, you notice that the *New York Times* and *USA Today,* whose weather maps are bounded by the precise, familiar outlines of the forty-eight contiguous United States, both assume that, as an American, you participate in the sunshine, clouds, and flurries of Salt Lake City, Miami, Fargo, Billings, and Los Angeles in a sense that you simply couldn't in those of a city fifty miles away across a national border. (In fact, *USA Today* goes even further in naturalizing the exclusion of Canada: since its snappy graphics give the weather map the illusion of projecting into three dimensions, a viewer of it receives the "information" that the North American continent dramatically drops off into the sea across the top of the United States.) What might surprise you is the powerfully familiarizing effect of nation-ness: that you could have failed to notice this before. In a sense, according to Anderson, these weather maps with all their apparent naturalness would be *the* defining icon of modern nationality. But the "same" experience at the Toronto airport turns out to be completely different. The *Globe and Mail,* and indeed every other Canadian newspaper I have seen, runs a weather map that extends southward at least as far as the Mason-Dixon Line in the United States. There is no presumption that the Toronto reader would identify more with the fog over Vancouver than with the blizzard approaching Detroit. Does this mean that Canada does not have "nation-ness"? Or instead that its nationness, having a different history from that of the United States, may well therefore have a structure different enough to put into question any single definition of the quality "nation-ness"?

The very blandness of the "American" compacting of borders—the, as it were, bad pun between the name of a continent and the name of a nation—how much must it not owe to the accidents of a history of geographic, economic, imperialist entitlement, a path into "nation-ness" no more "nor-

mal," no more *as opposed to* the same set or even the same *kinds* of definitional others, than that of the nation-ness of Canada, the different nation-
ness of Mexico, of the Philippines, of the Navajo Nation (within the United
States), of the Five Nations (across the U.S.-Canadian border), the nationalism of the non-nation Quebec, the non-nationalism of the non-nation
Hawaii, the histories of African American nationalisms, and so forth and so
forth and so forth. The "other" of *the* nation in a given political or historical
setting may be the prenational monarchy, the local ethnicity, the diaspora,
the transnational corporate, ideological, religious, or ethnic unit, the sub-
national locale or the ex-colonial, often contiguous unit; the colony may
become national vis-à-vis the homeland, or the homeland become national
vis-à-vis the nationalism of its colonies; the nationalism of the homeland
may be coextensive with or oppositional to its imperialism; and so forth.
Far beyond the pressure of crisis or exception, it may be that there exists for
nations, as for genders, simply no normal way to partake of the categorical
definitiveness of the national, no single kind of "other" of what a nation is
to which all can by the same structuration be definitionally opposed.

When it occurs to us, then, to run this question of national definition
athwart some already articulated questions of turn-of-the-century sexual
definition, we must be prepared to look in more directions at once than
one. For instance, as I have discussed at more length in *Epistemology of the
Closet*,[5] it is more than possible—it is almost unavoidable—to read *The
Picture of Dorian Gray* as an intensely and rather conventionally oriental-
izing text. This is played out in both the lavishly exhibitionistic openness,
and the techniques of coverture, of the novel's homosexual subject. It can
be seen, for instance, in the intertwining mutual camouflage and mutual
expression of the novel's gay plot with its drug addiction plot. The heavily
exoticized and glamorized opium commodity condenses many of the problematics of the natural vs. the unnatural/artificial, voluntarity vs. addiction,
and the domestic scene vs. the foreign substance, that seem in the first place
to have to do with the novel's modes of framing gay male identity. Yet the
very patency of Wilde's gay-affirming and gay-occluding Orientalism renders it difficult to turn back and see the outlines of the sexual body and the
national body sketched by his *occidentalism*. With Orientalism so ready-
to-hand a rubric for the relation to the Other, it is difficult to resist seeing
the desired English body, on the other hand, as simply the domestic Same.
Yet the sameness of this Same—or put another way, the *homo-* nature of
this sexuality—is no less open to question than the self-identicalness of the

5. Sedgwick, *Epistemology of the Closet*, chap. 3, from which this paragraph is drawn.

national borders of the domestic. After all, the question of the National in Wilde's own life only secondarily—though profoundly—involved the question of overseas empire in relation to European *patria.* To the contrary: Wilde, as an ambitious Irishman, and the son, intimate, and protégé of a celebrated Irish nationalist poet, can only have had as a fundamental element of his own sense of self an exquisitely exacerbated sensitivity to how by turns porous, brittle, elastic, chafing, embracing, exclusive, murderous, in every way contestable and contested were the membranes of "domestic" national definition signified by the ductile and elusive terms England, Britain, Ireland. Indeed, the consciousness of foundational and/or incipient national *difference* already internal to national *definition* must have been part of what Wilde literally embodied, in the expressive, specularized, and symptomatic relation in which he avowedly stood to his age. As a magus in the worship of the "slim rose-gilt soul"—the individual or generic figure of the "slim thing, gold-haired like an angel" that stood at the same time for a sexuality, a sensibility, a class, and a narrowly English national type— Wilde, whose own physical make was of an opposite sort and (in that context) an infinitely less appetizing, desirable, and placeable one, showed his usual uncanny courage ("his usual uncanny courage," *anglice* chutzpah) in foregrounding his own body so insistently as an index to such erotic and political meanings. The decades around the turn of the century marked the precipitous popularization, not only of the new word "homosexual," but of the very conception of male-male desire as a desire based on sameness. I have argued, also in *Epistemology of the Closet,* that a central trajectory in, for instance, *Dorian Gray,* is toward the establishment of men's love for men, no longer as a classicizing pedagogic/pederastic relation structured, like Basil's and Lord Henry's relations to Dorian, by diacritical differences between lover and beloved, but instead in the relatively modern terms, as slim rose-gilt Dorian's inescapably narcissistic mirror relation to his own figured body in the portrait. Wilde's alienizing physical heritage of unboundable bulk from his Irish nationalist mother, of a louche swarthiness from his Celticizing father, underlined with every self-foregrounding gesture of his person and persona the fragility, unlikelihood, and strangeness—at the same time, the transformative re-perceptualizing power—of the new "*homo-*" homosexual imagining of male-male desire. By the same pressure, it dramatized the uncouth nonequivalence of an English national body with a British with an Irish, as domestic grounds from which to launch a stable understanding of national/imperial relations.

Of course, Wilde's very hyper-indicativeness as a figure of his age means that there's no one else like him, but I do want to suggest that the mutual interrepresentations of emerging national and sexual definitions must be looked for at no less a level of complexity for other important figures, as well. Roger Casement's would be an obvious career in which to look for a heightening and contrastive braiding together of exoticizing British imperialist/anti-imperialist, with Irish nationalist, with homosexual identifications and identities. For him, as for Wilde, the question of the Other of a national, as of a sexual, identity was an irreducibly—and *sometimes* an enablingly—complex one. But even for someone like E. M. Forster whose national identity was in no sense a colonized one, the erotically expressive anti-imperialism of *A Passage to India* has as its other face the also anti-imperialistic, highly problematized English nationalism of *The Longest Journey,* whose shepherd nature-hero refuses an imperialist future in the colonies in favor of the homoerotically anthropomorphic, body-scaled and nationally figured landscape of his native valley. "I can't run up the [Druidic] Rings without getting tired, nor gallop a horse out of this view without tiring it, so what," he asks, "is the point of a boundless continent?" [6]

For Proust, on the other hand, whose plots of Dreyfusism and of gay recognition are the organizing principles for one another as they are for the volumes through which they ramify, the numinous identification of male homosexuality (which he figures in terms of a diaspora from the original homeland in Sodom) with a *pre*national, premodern, dynastic cosmopolitanism, through the figure of Charlus as much as through the Jews, is haunted by the specter of a sort of gay Zionism or pan-Germanism, a normalizing politics on the nominally ethnic model that would bring homosexual identity itself under the sway of what Nietzsche called "the *névrose nationale* with which Europe is sick." At the climax of his disquisition on "the men-women of Sodom," Proust explains,

> I have thought it as well to utter here a provisional warning against the lamentable error of proposing (just as people have encouraged a Zionist movement) to create a Sodomist movement and to rebuild Sodom. For, no sooner had they arrived there than the Sodomites would leave the town so as not to have the appearance of belonging to it, would take wives, keep mistresses in other cities where they would find, incidentally, every diversion that appealed to them. They would repair to Sodom only on days of supreme necessity, when their own town was empty, at those seasons when hunger drives the wolf from the woods.

6. Forster, *The Longest Journey,* p. 214.

In other words, everything would go on very much as it does to-day in London, Berlin, Rome, Petrograd, or Paris.[7]

And what finally to say of the virulently anti-German Nietzsche himself, with his homoerotic luxuriances of Europeanized sickness, the homo-sexual-panic-charged epistemologies of his political diagnostic techniques, his more than Proustian resistance to every nationalism as to every form of gay minority identity, and the more than Wildean passions of vicariousness with which he invested so rich a variety of male national bodies? I don't think any of these accounts will be simple ones to render—even to render visible. But we need to do so lest we continue to deal numbly around and along the eroticized borders of this apparently universal, factitiously time-less modern mapping of the national body.

7. Proust, *Remembrance of Things Past*, 2: 655–56.
"Nationalisms and Sexualities: As Opposed to What?" was written for a conference held at Harvard University in 1989. The essay is dedicated to Doris Sommer.

HOW TO BRING

YOUR KIDS UP GAY:

THE WAR ON

EFFEMINATE BOYS

In the summer of 1989 the U.S. Department of Health and Human Services released a study entitled "Report of the Secretary's Task Force on Youth Suicide." Written in response to the apparently burgeoning epidemic of suicides and suicide attempts by children and adolescents in the United States, the 110-page report contained a section analyzing the situation of gay and lesbian youth. It concluded that, because "gay youth face a hostile and condemning environment, verbal and physical abuse, and rejection and isolation from families and peers," young gays and lesbians are two to three times more likely than other young people to attempt and to commit suicide. The report recommends, modestly enough, an "end [to] discrimination against youths on the basis of such characteristics as . . . sexual orientation."

On 13 October 1989, Dr. Louis W. Sullivan, secretary of the Department of Health and Human Services, repudiated this section of the report—impugning not its accuracy, but, it seems, its very existence. In a written statement Sullivan said, "the views expressed in the paper entitled 'Gay Male and Lesbian Youth Suicide' do not in any way represent my personal beliefs or the policy of this Department. I am strongly committed to advancing traditional family values. . . . In my opinion, the views expressed in the paper run contrary to that aim." [1]

1. This information comes from reports in the *New York Native*, 23 September 1989, pp. 9–10; 13 November 1989, p. 14; 27 November 1989, p. 7.

It's always open season on gay kids. But where, in all this, are psycho-analysis and psychiatry? Where are the "helping professions"? In this dis-cussion of institutions, I mean to ask, not about Freud and the possibly spacious affordances of the mother-texts, but about psychoanalysis and psychiatry as they are functioning in the United States today.[2] I am espe-cially interested in revisionist psychoanalysis, including ego psychology, and in developments following on the American Psychiatric Association's much-publicized 1973 decision to drop the pathologizing diagnosis of homosexuality from its next Diagnostic and Statistical Manual (DSM-III). What is likely to be the fate of children brought under the influence of psychoanalysis and psychiatry today, post-DSM-III, on account of parents' or teachers' anxieties about their sexuality?

The monographic literature on the subject is, to begin with, as far as I can tell exclusively about boys. A representative example of this revision-ist, ego-based psychoanalytic theory would be Richard C. Friedman's *Male Homosexuality: A Contemporary Psychoanalytic Perspective,* pub-lished by Yale University Press in 1988.[3] (A sort of companion volume, though by a nonpsychoanalyst psychiatrist, is Richard Green's The 'Sissy Boy Syndrome' and the Development of Homosexuality [1987], also from Yale.)[4] Friedman's book, which lavishly acknowledges his wife and chil-dren, is strongly marked by his sympathetic involvement with the 1973 depathologizing movement. It contains several visibly admiring histories of gay men, many of them encountered in nontherapeutic contexts. These include "Luke, a forty-five-year-old career army officer and a life-long ex-clusively homosexual man" (RF, p. 152); and Tim, who was "burly, strong, and could work side by side with anyone at the most strenuous jobs": "gregarious and likeable," "an excellent athlete," Tim was "captain of [his high school] wrestling team and editor of the school newspaper" (pp. 206–7). Bob, another "well-integrated individual," "had regular sexual activity with a few different partners but never cruised or visited gay bars or baths. He did not belong to a gay organization. As an adult, Bob had had a

2. A particularly illuminating overview of psychoanalytic approaches to male homosexuality is available in Kenneth Lewes, *The Psychoanalytic Theory of Male Homosexuality* (New York: Penguin/NAL/Meridian, 1989).

3. Richard C. Friedman, *Male Homosexuality: A Contemporary Psychoanalytic Perspective* (New Haven, Conn.: Yale University Press, 1988). All citations will appear in parentheses in the text with RF.

4. Richard Green, *The "Sissy Boy Syndrome" and the Development of Homosexuality* (New Haven, Conn.: Yale University Press, 1987). Citations will appear in the text with RG.

stable, productive work history. He had loyal, caring, durable friendships with both men and women" (pp. 92–93). Friedman also, by way of comparison, gives an example of a *hetero*sexual man with what he considers a highly integrated personality, who happens to be a combat jet pilot: "Fit and trim, in his late twenties, he had the quietly commanding style of an effective decision maker" (p. 86).[5]

Is a pattern emerging? Revisionist analysts seem prepared to like some gay men, but the healthy homosexual is one who (a) is already grown up, and (b) acts masculine. In fact, Friedman correlates, in so many words, adult gay male effeminacy with "global character pathology" and what he calls "the lower part of the psychostructural spectrum" (p. 93). In the obligatory paragraphs of his book concerning "the question of when behavioral deviation from a defined norm should be considered psychopathology," Friedman makes explicit that, while "clinical concepts are often somewhat imprecise and admittedly fail to do justice to the rich variability of human behavior," a certain baseline concept of pathology will be maintained in his study, and that that baseline will be drawn in a very particular place. "The distinction between nonconformists and people with psychopathology is usually clear enough during childhood. Extremely and chronically effeminate boys, for example, should be understood as falling into the latter category" (pp. 32–33).

"For example," "extremely and chronically effeminate boys"—this is the abject that haunts revisionist psychoanalysis. The same DSM-III that, published in 1980, was the first that did not contain an entry for "homosexuality," was also the first that *did* contain a new diagnosis, numbered (for insurance purposes) 302.60: "Gender Identity Disorder of Childhood." Nominally gender-neutral, this diagnosis is actually highly differential between boys and girls: a girl gets this pathologizing label only in the rare case of asserting that she actually is anatomically male (e.g., "that she has, or will grow, a penis"); while a boy can be treated for Gender Identity Disorder of Childhood if he merely asserts "that it would be better not to have a penis"—*or* alternatively, if he displays a "preoccupation with female stereotypical activities as manifested by a preference for either cross-dressing or simulating female attire, or by a compelling desire to par-

5. It is worth noting that the gay men Friedman admires always have completely discretionary control over everyone else's knowledge of their sexuality; no sense that others may have their own intuitions that they are gay; no sense of physical effeminacy; no visible participation in gay (physical, cultural, sartorial) semiotics or community. For many contemporary gay people, such an existence would be impossible; for a great many, it would seem starvingly impoverished in terms of culture, community, and meaning.

ticipate in the games and pastimes of girls."[6] While the decision to remove "homosexuality" from DSM-III was a highly polemicized and public one, accomplished only under intense pressure from gay activists outside the profession, the addition to DSM-III of "Gender Identity Disorder of Child-hood" appears to have attracted no outside attention at all—nor even to have been perceived as part of the same conceptual shift.[7]

Indeed, the gay movement has never been quick to attend to issues concerning effeminate boys. There is a discreditable reason for this in the marginal or stigmatized position to which even adult men who are effeminate have often been relegated in the movement.[8] A more understandable reason than effeminophobia, however, is the conceptual need of the gay movement to interrupt a long tradition of viewing gender and sexuality as continuous and collapsible categories—a tradition of assuming that anyone, male or female, who desires a man must by definition be feminine; and that anyone, male or female, who desires a woman must by the same token be masculine. That one woman, *as a woman,* might desire another; that one man, *as a man,* might desire another: the indispensable need to make these powerful, subversive assertions has seemed, perhaps, to require a relative deemphasis of the links between gay adults and gender-nonconforming children. To begin to theorize gender and sexuality as distinct though intimately entangled axes of analysis has been, indeed, a great advance of recent lesbian and gay thought.

There is a danger, however, that that advance may leave the effeminate boy once more in the position of the haunting abject—this time the haunting abject of gay thought itself. This is an especially horrifying possibility if—as many studies launched from many different theoretical and political positions have suggested—for any given adult gay man, wherever he may be at present on a scale of self-perceived or socially ascribed mascu-

6. *Diagnostic and Statistical Manual of Mental Disorders* 3d ed. (Washington, D.C.: American Psychiatric Association, 1980), pp. 265–66.

7. The exception to this generalization is Lawrence Mass, whose *Dialogues of the Sexual Revolution,* vol. 1, *Homosexuality and Sexuality* (New York: Harrington Park Press, 1990) collects a decade's worth of interviews with psychiatrists and sex researchers, originally conducted for and published in the gay press. In these often illuminating interviews, a number of Mass's questions are asked under the premise that "American psychiatry is simply engaged in a long, subtle process of reconceptualizing homosexuality as a mental illness with another name—the 'gender identity disorder of childhood'" (p. 214).

8. That relegation may be diminishing as, in many places, "queer" politics come to overlap and/or compete with "gay" politics. Part of what I understand to be the exciting charge of the very word "queer" is that it embraces, instead of repudiating, what have for many of us been formative childhood experiences of difference and stigmatization.

linity (ranging from extremely masculine to extremely feminine), the like-lihood is disproportionately high that he will have a childhood history of self-perceived effeminacy, femininity, or nonmasculinity.[9] In this case the eclipse of the effeminate boy from adult gay discourse would represent more than a damaging theoretical gap; it would represent a node of annihilating homophobic, gynephobic, and pedophobic hatred internalized and made central to gay-affirmative analysis. The effeminate boy would come to function as the discrediting open secret of many politicized adult gay men.

One of the most interesting aspects—and by interesting I mean cautionary—of the new psychoanalytic developments is that they are based on *precisely* the theoretical move of distinguishing gender from sexuality. This is how it happens that the *de*pathologization of an atypical sexual object-choice can be yoked to the *new* pathologization of an atypical gender identification. Integrating the gender-constructivist research of, for example, John Money and Robert Stoller, research that many have taken (though perhaps wrongly) as having potential for feminist uses, this work posits the very early consolidation of something called Core Gender Identity—one's basal sense of being male or female—as a separate stage prior to, even conceivably independent of, any crystallization of sexual fantasy or sexual object choice. Gender Identity Disorder of Childhood is seen as a pathology involving the Core Gender Identity (failure to develop a CGI consistent with one's biological sex); sexual object-choice, on the other hand, is unbundled from this Core Gender Identity through a reasonably space-making series of two-phase narrative moves. Under the pressure, ironically, of having to show how gay adults whom he considers well-integrated personalities do sometimes evolve from children seen as the very definition of psychopathology, Friedman unpacks several developmental steps that have often otherwise been seen as rigidly unitary.[10]

9. For descriptions of this literature, see Friedman, *Male Homosexuality,* pp. 33–48; and Green, *The "Sissy Boy Syndrome,"* pp. 370–90. The most credible of these studies from a gay-affirmative standpoint would be A. P. Bell, M. S. Weinberg, and S. K. Hammersmith, *Sexual Preference: Its Development in Men and Women* (Bloomington: Indiana University Press, 1981), which concludes: "Childhood Gender Nonconformity turned out to be more strongly connected to adult homosexuality than was any other variable in the study" (p. 80).

10. Priding himself on his interdisciplinarity, moreover, he is much taken with recent neuroendocrinological work suggesting that prenatal stress on the mother may affect structuration of the fetal brain in such a way that hormonal cues to the child as late as adolescence may be processed differentially. His treatment of these data as data is neither very responsible (e.g., problematical results that point only to "hypothetical differences" in one chapter (p. 24) have been silently upgraded to positive "knowledge" two chapters later (p. 51)) nor very impartial (for instance, the conditions hypothesized as conducing to gay development are invariably re-

One serious problem with this way of distinguishing between gender and sexuality is that, while denaturalizing sexual object-choice, it radically *re*naturalizes gender. All ego psychology is prone, in the first place, to structuring developmental narrative around a none-too-dialectical trope of progressive *consolidation* of self. To place a very early core-gender determinant (however little biologized it may be) at the center of that process of consolidation seems to mean, essentially, that for a nontranssexual person with a penis, nothing can ever be assimilated to the self through this process of consolidation unless it can be assimilated *as masculinity.* For even the most feminine-self-identified boys, Friedman uses the phrases "sense of masculine self-regard" (RF, p. 245), "masculine competency" (p. 20), and "self-evaluation as appropriately masculine" (p. 244) as synonyms for any self-esteem and, ultimately, for any *self.* As he describes the interactive process that leads to any ego consolidation in a boy:

> Boys measure themselves in relation to others whom they estimate to be similar. [For Friedman, this means only men and other boys.] Similarity of self-assessment depends on consensual validation. The others must agree that the boy is and will remain similar to them. The boy must also view both groups of males (peers and older men) as appropriate for idealization. Not only must he be like them in some ways, he must want to be like them in others. They in turn must want him to be like them. Unconsciously, they must have the capacity to identify with him. This naturally occurring [!] fit between the male social world and the boy's inner object world is the juvenile phase-specific counterpoint to the preoedipal child's relationship with the mother. (p. 237)

The reason effeminate boys turn out gay, according to this account, is that other men don't validate them as masculine. There is a persistent, wistful fantasy in this book: "One cannot help but wonder how these [pre-homosexual boys] would have developed if the males they idealized had had a more flexible and abstract sense of masculine competency" (p. 20).

---

ferred to as *inadequate* androgenization (p. 14), *deficit* (p. 15), etc.). But his infatuation with this model does have two useful effects. First, it seems to generate by direct analogy this further series of two-phase narratives about psychic development, narratives that discriminate between the circumstances under which a particular psychic structure is *organized* and those under which it is *activated,* that may turn out to enable some new sinuosities for other, more gay-embracing and pluralist projects of developmental narration. (This analogical process is made explicit on pp. 241–45.) And second, it goes a long way toward detotalizing, demystifying, and narrativizing in a recognizable way any reader's sense of the threat (the promise?) presented by a supposed neurobiological vision of the already-gay male body.

For Friedman, the increasing flexibility in what kinds of attributes or activities *can* be processed as masculine, with increasing maturity, seems fully to account for the fact that so many "gender-disturbed" (effeminate) little boys manage to grow up into "healthy" (masculine) men, albeit after the phase where their sexuality has differentiated as gay.

Or rather, it *almost* fully accounts for it. There is a residue of mystery, resurfacing at several points in the book, about why most gay men turn out so resilient—about how they even survive—given the profound initial deficit of "masculine self-regard" characteristic of many proto-gay childhoods, and the late and relatively superficial remediation of it that comes with increasing maturity. Given that "the virulence and chronicity of [social] stress [against it] puts homosexuality in a unique position in the human behavioral repertoire," how to account for "the fact that severe, persistent morbidity does not occur more frequently" among gay adolescents (RF, p. 205)? Friedman essentially throws up his hands at these moments. "A number of possible explanations arise, but one seems particularly likely to me: namely, that homosexuality is associated with some psychological mechanism, not understood or even studied to date, that protects the individual from diverse psychiatric disorders" (p. 236). It "might include mechanisms influencing ego resiliency, growth potential, and the capacity to form intimate relationships" (p. 205). And "it is possible that, for reasons that have not yet been well described, [gender-disturbed boys'] mechanisms for coping with anguish and adversity are unusually effective" (p. 201).

These are huge blank spaces to be left in what purports to be a developmental account of proto-gay children. But given that ego-syntonic consolidation for a boy can come only in the form of masculinity, given that masculinity can be conferred only by men (p. 20), and given that femininity, in a person with a penis, can represent nothing but deficit and disorder, the one explanation that could *never* be broached is that these mysterious skills of survival, filiation, and resistance could derive from a secure identification with the resource richness of a mother. Mothers, indeed, have nothing to contribute to this process of masculine validation, and women are reduced in the light of its urgency to a null set: any involvement in it by a woman is overinvolvement, any protectiveness is overprotectiveness, and, for instance, mothers "proud of their sons' nonviolent qualities" are manifesting unmistakable "family pathology" (p. 193).

For both Friedman and Green, then, the first, imperative developmental task of a male child or his parents and caretakers is to get a properly male Core Gender Identity in place as a basis for further and perhaps more flexible explorations of what it may be to *be* masculine—i.e., for a male person, to be *human*. Friedman is rather equivocal about whether this masculine CGI necessarily entails any particular content, or whether it is an almost purely formal, preconditional differentiation that, once firmly in place, can cover an almost infinite range of behaviors and attitudes. He certainly does not see a necessary connection between masculinity and any scapegoating of male homosexuality; since ego psychology treats the development of male heterosexuality as nonproblematical after adolescence, as not involving the suppression of any homosexual or bisexual possibility (pp. 263–67), and therefore as completely unimplicated with homosexual panic (p. 178), it seems merely an unfortunate, perhaps rectifiable misunderstanding that for a proto-gay child to identify "masculinely" might involve his identification with his own erasure.

The renaturalization and enforcement of gender assignment is not the worst news about the new psychiatry of gay acceptance, however. The worst is that it not only fails to offer, but seems conceptually incapable of offering, even the slightest resistance to the wish endemic in the culture surrounding and supporting it: the wish that gay people *not exist*. There are many people in the worlds we inhabit, and these psychiatrists are unmistakably among them, who have a strong interest in the dignified treatment of any gay people who may happen already to exist. But the number of persons or institutions by whom the existence of gay people is treated as a precious desideratum, a needed condition of life, is small. The presiding asymmetry of value assignment between hetero and homo goes unchallenged everywhere: advice on how to help your kids turn out gay, not to mention your students, your parishioners, your therapy clients, or your military subordinates, is less ubiquitous than you might think. On the other hand, the scope of institutions whose programmatic undertaking is to prevent the development of gay people is unimaginably large. There is no major institutionalized discourse that offers a firm resistance to that undertaking: in the United States, at any rate, most sites of the state, the military, education, law, penal institutions, the church, medicine, and mass culture enforce it all but unquestioningly, and with little hesitation at even the recourse to invasive violence.

These books, and the associated therapeutic strategies and institutions, are not about invasive violence. What they are about is a train of squalid lies. The overarching lie is the lie that they are predicated on anything but the therapists' disavowed desire for a nongay outcome. Friedman, for instance, speculates wistfully that—with proper therapeutic intervention—the sexual orientation of one gay man whom he describes as quite healthy might conceivably (not have *been changed* but) "have shifted *on its own*" (Friedman's italics): a speculation, he artlessly remarks, "not value-laden with regard to sexual orientation" (p. 212). Green's book, composed largely of interview transcripts, is a tissue of his lies to children about their parents' motives for bringing them in. (It was "not to prevent you from becoming homosexual," he tells one young man who had been subjected to behavior modification, "it was because you were unhappy" (RG, p. 318); but later on the very same page, he unself-consciously confirms to his trusted reader that "parents of sons who entered therapy were . . . worried that the cross-gender behavior portended problems with later sexuality.") He encourages predominantly gay young men to "reassure" their parents that they are "bisexual" ("Tell him just enough so he feels better" [RG, p. 207]) and to consider favorably the option of marrying and keeping their wives in the dark about their sexual activities (p. 205). He lies to himself and to us in encouraging patients to lie to him. In a series of interviews with Kyle, for instance, the boy subjected to behavioral therapy, Green reports him as saying that he is unusually withdrawn—" 'I suppose I've been overly sensitive when guys look at me or something ever since I can remember, you know, after my mom told me why I have to go to UCLA because they were afraid I'd turn into a homosexual' " (p. 307); as saying that homosexuality "is pretty bad, and I don't think they should be around to influence children. . . . I don't think they should be hurt by society or anything like that—especially in New York. You have them who are into leather and stuff like that. I mean, I think that is really sick, and I think that maybe they should be put away" (p. 307); as saying that he wants to commit violence on men who look at him (p. 307); and as saying that if he had a child like himself, he "would take him where he would be helped" (p. 317). The very image of serene self-acceptance?

Green's summary:

> Opponents of therapy have argued that intervention underscores the child's "deviance," renders him ashamed of who he is, and makes him suppress his "true self." Data on psychological tests do not support this contention; nor does the content of clinical interviews. The boys look back favorably on treatment. They would endorse such intervention if they were the father of a "feminine" boy. Their reason is to reduce childhood conflict and social stigma. Therapy with these boys appeared to accomplish this. (p. 319)

Consistent with this, Green is obscenely eager to convince parents that their hatred and rage at their effeminate sons is really only a desire to protect them from peer-group cruelty—even when the parents name *their own* feelings as hatred and rage (pp. 391–92). Even when fully one-quarter of parents of gay sons are *so* interested in protecting them from social cruelty that, when the boys fail to change, their parents kick them out on the street! Green is withering about mothers who display any tolerance of their sons' cross-gender behavior (pp. 373–75). In fact, his bottom-line identifications as a clinician actually seem to lie with the enforcing peer group: he refers approvingly at one point to "therapy, be it formal (delivered by paid professionals) or informal (delivered by the peer group and the larger society via teasing and sex-role standards)" (p. 388).

Referring blandly on one page to "psychological intervention directed at increasing [effeminate boys'] comfort with being male" (p. 259), Green says much more candidly on the next page, "the rights of parents to oversee the development of children is a long-established principle. Who is to dictate that parents may not try to raise their children in a manner that maximizes the possibility of a heterosexual outcome?" (p. 260). Who indeed—if the members of this profession can't stop seeing the prevention of gay people as an ethical use of their skills?

Even outside the mental health professions and within more authentically gay-affirmative discourses, the theoretical space for supporting gay development is, as I've pointed out in the introduction to *Epistemology of the Closet*, narrow. Constructivist arguments have tended to keep hands off the experience of gay and proto-gay kids. For gay and gay-loving people, even though the space of cultural malleability is the only conceivable theater for our effective politics, every step of this constructivist nature/culture argument holds danger: the danger of the difficulty of intervening in the seemingly natural trajectory from identifying a place of cultural malleability, to inventing an ethical or therapeutic mandate for cultural manipulation, to the overarching, hygienic Western fantasy of a world without any more homosexuals in it.

That's one set of dangers, and it is as against them, as I've argued, that essentialist and biologizing understandings of sexual identity accrue a certain gravity. The resistance that seems to be offered by conceptualizing an unalterably *homosexual body,* to the social-engineering momentum apparently built into every one of the human sciences of the West, can reassure profoundly. At the same time, however, in the postmodern era it is becoming increasingly problematical to assume that grounding an identity in biology or "essential nature" is a stable way of insulating it from societal interference. If anything, the gestalt of assumptions that undergirds nature/nurture debates may be in process of direct reversal. Increasingly it is the conjecture that a particular trait is genetically or biologically

based, *not* that it is "only cultural," that seems to trigger an estrus of manipulative fantasy in the technological institutions of the culture. A relative depressiveness about the efficacy of social-engineering techniques, a
high mania about biological control: the Cartesian bipolar psychosis that
always underlay the nature/nurture debates has switched its polar assignments without surrendering a bit of its hold over the collective life. And
in this unstable context, the dependence on a specified *homosexual body*
to offer resistance to any gay-eradicating momentum is tremblingly vulnerable. AIDS, though it is used to proffer every single day to the news-
consuming public the crystallized vision of a world after the homosexual,
could never by itself bring about such a world. What whets these fantasies more dangerously, because more blandly, is the presentation, often in
ostensibly or authentically gay-affirmative contexts, of biologically based
"explanations" for deviant behavior that are absolutely invariably couched
in terms of "excess," "deficiency," or "imbalance"—whether in the hor-
mones, in the genetic material, or, as is currently fashionable, in the fetal
endocrine environment. If I had ever, in any medium, seen any researcher or
popularizer refer even once to any supposed gay-producing circumstance
as the *proper* hormone balance, or the *conducive* endocrine environment,
for gay generation, I would be less chilled by the breezes of all this technological confidence. As things are, a medicalized dream of the prevention of
gay bodies seems to be the less visible, far more respectable underside of
the AIDS-fueled public dream of their extirpation.

In this unstable balance of assumptions between nature and culture, at
any rate, under the overarching, relatively unchallenged aegis of a culture's
desire that gay people *not be,* there is no unthreatened, unthreatening theoretical home for a concept of gay and lesbian origins. What the books I
have been discussing, and the institutions to which they are attached, demonstrate is that the wish for the dignified treatment of already-gay people
is necessarily destined to turn into either trivializing apologetics or, much
worse, a silkily camouflaged complicity in oppression—in the absence of a
strong, explicit, *erotically invested* affirmation of some people's felt desire
or need that there be gay people in the immediate world.

"How to Bring Your Kids Up Gay" was written in 1989 for a Modern Language Association
panel. Jack Cameron pointed me in the direction of the texts discussed here, and Cindy Patton fortified my resistance to them.

ACROSS

GENDERS,

ACROSS

SEXUALITIES

WILLA

CATHER

AND

OTHERS

Willa Cather's 1905 short story, "Paul's Case: A Study in Temperament," which as its title might suggest sometimes reads like a case study from Richard Green's book on the "sissy boy syndrome," was Cather's own life-long favorite among her stories, the one she re-published most and the only one she allowed to be anthologized by others. The *omphalos* of her continuing attachment to this story is oddly difficult to locate; but (if it's permissible to complicate the navel metaphor in this way) the knotted-up surgical scar of her *de*tachment from the story's main character is almost its first legible sign. "It was Paul's afternoon to appear before the faculty of the Pittsburgh High School to account for his various misdemeanors," the story begins; with a "rancor and aggrievedness" for which they themselves feel shamed by their inability to account, the teachers of this tense, unlovely, effeminate, histrionic boy "fell upon him without mercy, his English teacher leading the pack."[1]

Once, when he had been making a synopsis of a paragraph at the blackboard, his English teacher had stepped to his side and attempted to guide his hand. Paul had started back with a shudder and thrust his hands violently behind him. The astonished woman could scarcely have been more hurt and embarrassed had he struck at her. The insult was so involuntary and definitely personal as to be unforgettable. In one way and another, he

---

1. Willa Cather, "Paul's Case," *Five Stories* (New York: Vintage Books, 1956), p. 149. Further citations will appear in parentheses in the text.

had made all his teachers, men and women alike, conscious of the same feeling of physical aversion. (p. 150)

The equivocalness of "the same feeling of physical aversion"—does the aversion live in his body, or in theirs; or is the aversiveness tied up with a certain threat of de-differentiation between them?—seems fulfilled when it is the teachers, and not the object of their discipline, who "left the building dissatisfied and unhappy; humiliated to have felt so vindictive toward a mere boy, to have uttered this feeling in cutting terms, and to have set each other on, as it were, in the grewsome game of intemperate reproach" (p. 152). And that evening, when the English teacher arrives at a concert for which the boy is officiously acting as usher, it is she who betrays "some embarrassment when she handed Paul the tickets, and a *hauteur* which subsequently made her feel very foolish" (p. 153).

The Pittsburgh High School English teacher evaporates as a character from "Paul's Case" shortly thereafter—"Paul forgot even the nastiness of his teacher's being there" (p. 154). There is nothing in the story to suggest that her unushered homeward steps will take her, though they would have taken one Pittsburgh high school English teacher any evening in 1905, back to a bedroom shared with a sumptuously beautiful young woman, Isabelle McClung, who has defied her parents to the extent of bringing her imposing lover, Willa Cather, into the family home to live. While this English teacher doesn't require to be identified with that English teacher, however, it is also less easy than it might be to differentiate firmly their attitudes toward what Katherine Anne Porter confidently labels as "a real 'case' in the clinical sense of the word," that is to say, "boys like Paul."[2]

Paul's teachers feel humiliated because they have found themselves momentarily unified in a ritual of scapegoating, without being at all clear what it is in the scapegoat that deserves torment or even what provokes this sudden communal structuration. Cather herself had some history of being an effeminophobic bully; perhaps also of feeling shamed by being one. Ten years before "Paul's Case," for instance, in 1895, the year when a new homophobic politics of indignity had its watershed international premiere in the trials of Oscar Wilde, Cather, as a young journalist, published two columns on Wilde. In each of these, she harnesses her own prose eagerly to the accelerating rhetoric of the public auto-da-fé. By her account Wilde is not just some random sinner—certainly not the object of any injustice—but some-

2. Katherine Anne Porter, "A Note," in Willa Cather, *The Troll Garden* (New York: New American Library, 1971), pp. 150–51.

thing more, a signal criminal of Luciferian stature, the "ghastly eruption that makes [society] hide its face in shame." [3] Wilde "is in prison now, most deservedly so. Upon his head is heaped the deepest infamy and the darkest shame of his generation. Civilization shudders at his name, and there is absolutely no spot on earth where this man can live. Cain's curse was light compared with his" (BS, p. 392). Joining so perfervidly in the public scapegoating of Wilde, Cather also seems, however, to assert a right, earned by the very excess of her revilement, to define for herself, and differently from the way the society or for that matter the courts had defined it, the "true" nature of Wilde's sin and hence the true justification for his punishment of two years' imprisonment with hard labor. Wilde's disgrace is, in Cather's account, no isolated incident but "the beginning of a national expiation" for "the sin which insults the dignity of man, and of God in whose image he was made," "the potentiality of all sin, the begetter of all evil" (p. 390). And that sin? "—Insincerity" (p. 390). "Art that is artificial and insincere" is the true "sin against the holy spirit," "for which there is no forgiveness in Heaven, no forgetting in Hell" (p. 391). The odd oversight of the framers of the Criminal Law Amendment Act and the 1885 Labouchère Amendment to it, in omitting to include by name the unspeakable crimes of artificiality and insincerity, doesn't slow Cather a bit in this determined act of redefinition. Reactivating the ancient, barely latent definitional antithesis between homosexual acts and *the natural,* Cather strongly reinforces the assaultive received association between Wilde's sexuality and a reprobated, putatively feminine love of artifice. At the same time, though, distinguishing however slightly and invidiously the one crime from the other ("The sins of the body are very small compared with that" [p. 391]), she also holds open a small shy gap of nonidentification between the two in which some nascent germ of gay-affirmative detachment, of critique, or even of outlaw-love might shelter to await her own less terroristic or terrorized season.

Had that season arrived with the writing a decade later of "Paul's Case"? In the early, Pittsburgh part of the story, it seems, to the contrary, that the identification between Paul's pathology on the one hand and his insincerity and artificiality on the other is so seamless that the former is to be fully evoked by the latter, and through a mercilessly specular, fixated point of view that takes his theatrical self-presentation spitefully at its word. "His eyes were remarkable for a certain hysterical brilliancy, and he continu-

3. Willa Cather, *The Kingdom of Art: Willa Cather's First Principles and Critical Statements, 1893–1896,* Bernice Slote, ed. (Lincoln: University of Nebraska Press, 1966), p. 390. Cited in text as "BS."

ally used them in a conscious, theatrical sort of way, peculiarly offensive in a boy" (pp. 149–50). Paul's glance is a jerkily unsteady one—"Paul was always smiling, always glancing about him, seeming to feel that people might be watching him and trying to detect something" (p. 151)—but the gaze of the narrative at him is so unresting as to give point to his desperate way of regarding the world; one English teacher, at least, is eternally there to describe him "looking wildly behind him now and then to see whether some of his teachers were not there to witness his lightheartedness" (p. 152). Like Cather's Wilde a decade earlier, it seems as if Paul is to be hounded to exhaustion or death for a crime that hovers indeterminately between sex/gender irregularity on the one hand and, on the other, spoilt sensibility or bad art. The invidious need of a passionate young lesbian to place, and at a distance, the lurid, contagious scandal of male homosexuality: it is as if that were not quite to be disentangled from the invidious need of a hungry young talent to distinguish itself once and for all from the "hysterical" artifice of the hapless youth who needs talent but hasn't it.

If the early parts of "Paul's Case" seem written from the unloving compulsions of the English teacher, however, the latter part of the story, after Paul has stolen a thousand dollars from his father's employer and run away to New York, opens out to what seems to me an amazing tenderness of affirmation. How common is it, in a fictional tradition ruled by *le mensonge romantique*, for a powerfully desiring character to get the thing that he desires, and to learn immediately that he was right—that what he wanted really is the thing that would make him happy? And especially in the specific tradition of, shall we say, *Madame Bovary*, the fictional lineage whose geography consists solely of *the provinces* and *the capital*, and whose motive for desire is an acculturated stimulus that is explicitly said to be at once entire artifice and yet less than art? But this is what happens to the furtive, narrow-chested, nerve-twanging Paul in snowy New York. Furnished in one single expert shopping spree with suits, shoes, hat, scarf pin, brushes, handsome luggage, installed in a suite in the Waldorf, catered to with violets and jonquils, bathed, warmed, rested, and dressed, the opera in prospect and the popping of champagne corks to waltz music, on Paul's looking in the mirror the next day, "Everything was quite perfect; he was exactly the kind of boy he had always wanted to be" (p. 167). And it seems to be true. Even when, days later, running out of money and time he throws himself in front of a train and is crushed to atoms, nothing has undermined the preciousness for Paul of, not art, but the chrism of a feminized or homosexualized culture and artifice. "He felt now that his surroundings explained him" (p. 169); the lights forgivingly, expensively lowered, himself at last the seeing consciousness of the story, even the past of his life and his body are, as in coming out, re-knit with the new, authoritative fingers of his own eyes. "He realized well enough that he had always been tormented by . . . a sort of apprehensive dread that, of late years, as the meshes of the lies he had told closed about him, had been pulling the muscles of his body tighter and tighter. . . . There had always been the shadowed corner, the dark place into which he dared not look, but from which something seemed always to be watching him" (p. 166). But now, "these were his people, he told himself" (p. 168). "It would be impossible for anyone to humiliate him" (p. 169). If Cather, in this story, does something to cleanse her own sexual body of the carrion stench of Wilde's victimization, it is thus (unexpectedly) by identifying with what seems to be Paul's sexuality not in spite of but *through* its saving reabsorption in a gender-liminal (and very specifically classed) artifice that represents at once a particular subculture and culture itself.

Cather's implicit reading here of the gendering of sex picks out *one* possible path through the mazed junction at which long-residual issues of gender and class definition intersected new turn-of-the-century mappings of sexual choice and identification. For enduringly since at least the turn of the century, there have presided two contradictory tropes of gender through which same-sex desire could be understood. I think of these as gender liminality and gender separatism. On the one hand, there was, and there persists differently coded (in the homophobic folklore and science around the "sissy boy" and his mannish sister, but also in the heart and guts of much living gay and lesbian culture) the trope of inversion, "a woman's soul trapped in a man's body" and vice versa. As writers such as Christopher Craft have made clear, one vital impulse of this trope is the preservation of an essential *heterosexuality* of desire itself, through a particular reading of the homosexuality of persons: desire, in this view, by definition subsists in the current that runs between one male self and one female self, in whatever sex of bodies these selves may be manifested. Charged as it has been with powerful meanings both gay-loving and gay-hating, the persistence of the inversion or liminality trope has been yoked, however, to that of its contradictory counterpart, the trope of gender separatism. As the gender-separatist substitution of the concept of the "woman-identified woman" for the gender-liminal stereotype of the mannish lesbian suggests, as indeed does the concept of the continuum of male or female homosocial desire, this trope tends to reassimilate to one another identification and desire, where inversion models, by contrast, depend conceptually on their distinctness. Gender-separatist models would thus place the woman-loving woman and the man-loving man each at the "natural" defining center of their own gender, again in contrast to inversion models that locate gay people—whether biologically or culturally—at the threshold between genders.[4]

In what I am reading as Cather's move in "Paul's Case," the mannish lesbian author's coming together with the effeminate boy on the ground of a certain distinctive position of gender liminality is also a move toward a minority gay identity whose more effectual cleavage, whose more determining separatism, would be that of homo/hetero *sexual* choice rather than that of male/female *gender*. The playing out of this story through the intertwinings of such gender-polarized terms as culture, artifice, and the punitive gaze, however, means that the thick semantics of gender asymmetry will cling to the syntax however airy of gender-crossing and recrossing—clogging or rendering liable to slippage the gears of reader or authorial relation with a special insistence of viscosity.

Such formulations as these might push us toward some new hypotheses and new clusterings-together, in the rich tradition of cross-gender inventions of homosexuality of the past century whose distinguished constituency might include—along with Cather—Henry James, Proust, Yourcenar, Compton-Burnett, Renault, and more recent writers in sometimes less literary forms such as Gayle Rubin, Esther Newton, Susan Sontag, Judy Grahn, Joanna Russ—or, I might add, Prince in his astonishingly Proustian hit single "If I Was Your Girlfriend." We could ask, for instance, about a text like James's *The Bostonians,* whether certain vindictive wrenchings of it out of "shape," warpings in its illusionistic surface of authorial control and address, might not represent less a static parti pris *against* women's desire for women than a dangerously unresolved question *about* it. How far, the novel asks—or more powerfully resists asking for fear of either the answer or no answer or too many answers—how far are these two things parallel or comparable: the ventriloquistic, half-contemptuous, hot desire of Olive Chancellor for a girl like Verena Tarrant; the ventriloquistic, half-contemptuous, hot desire of Henry James for a boy like Basil Ransome? To the degree that they aren't parallel, the intimate access of the authorial consciousness to the characters' must be dangerously compromised; to the degree that they are, so must its panoptic framing authority of distance and diagnostic privilege. Whether (in those waning decades of an intensive American homosociality) men's desire for men represents most the central and naturalizing maleness of its communicants, and women's for women an ultra-femininity unnatural only for its overtypifying concentration; or whether (in those inaugural years of medical, psychiatric,

4. This analysis comes from my *Epistemology of the Closet,* pp. 86–90.

and legal discourses, both pro- and antigay, that variously distinguished "homosexuals" male and female as a singular minority) a same-sex desire necessarily invoked, as well, a *cross*-gender liminality; the criss-crossed reading of such desires so ruptures the authorial surface of James's writing that what shoves through it here is the fist not of a male-erotic *écriture* but less daringly of a woman-hating and feminist-baiting violence of panic.

A more truly sexy moment, in Proust, gives more the thrill of authorial availability through these eruptive diagonals of definition. After the flight and death of his probably lesbian mistress Albertine, the narrator of *A la recherche*, his jealousy only the more inflamed by its posthumousness, dispatches on an errand of detection and reconstruction along the trail of her elopement a functionary named Aimé, a headwaiter who has himself already been both a procurer and an object of desire both within and across gender through several volumes of this novel. Aimé sends back an account of Albertine which the narrator reproduces thus, its peculiarity being that Aimé, half-educated, "when he meant to put inverted commas . . . put brackets, and when he meant to put something in brackets . . . put it in inverted commas":[5]

> "The young laundry-girl confessed to me that she enjoyed playing around with her girlfriends and that seeing Mlle Albertine was always rubbing up against her in her bathing-wrap she made her take it off and used to caress her with the tongue along her throat and arms. . . . After that she told me nothing more, but being always at your service and ready to do anything to oblige you, I took the young laundry-girl to bed with me. She asked me if I would like her to do to me what she used to do to Mlle Albertine when she took off her bathing-dress. And she said to me: (If you could have seen how she used to wriggle, that young lady, she said to me (oh, it's too heavenly) and she got so excited that she could not keep from biting me.) I could still see the marks on the laundry-girl's arms."[6]

The deroutinization of the subordinating work of punctuation here strips away the insulation of the text aginst every juxtaposition of sexualities. Albertine to the laundry girl, the laundry girl to Aimé, Aimé to the male narrator, the narrator to the reader: the insubordinated address of pain and ecstasy is anchored only at the last moment by its cryptic laundry mark, residue of a rinsing and ravenous illegible rapture.

5. Proust, *Remembrance of Things Past*, vol. 3, p. 525.
6. Ibid., vol.3, p. 535.

This highly chiasmic organization of homo/heterosexual definition in Proust, which prompts Leo Bersani, for instance, to formulate "the ontological necessity of homosexuality [in the other sex] in a kind of universal *hetero*sexual relation of all human subjects to their own desires,"[7] confirms that insofar as a growing minority gay identity does or did depend on a model of gender liminality, it tends to invest with meaning transpositions between *male* and *female* homosexuality, but also between the bonds of *homo*- and *hetero*sexuality themselves. If the Proustian example suggests that this choreography of crossings, identifications, and momentary symmetries, this hummingbird ballet of within and without the parenthesis, represents a utopian possibility somewhere to one side of the stresses of gender or of other exploitations, however, I think there is evidence that the truth of this choreography of cross-translation is at once less blithe and more interesting.

Back, for one final example, to Willa Cather: to her beautiful and difficult novel, *The Professor's House,* whose eponymous domesticity (the house itself, that is to say, not the Professor) is biographically thought to allude to the enabling provision for Cather's own writing of a room of her own, first by Isabelle McClung and then perhaps by her more domesticated and serviceable companion/lover of decades, Edith Lewis. Though the love that sustains these necessary facilitations has been a lesbian one, however, its two crystallizations in the novel are both cross-translations: one across gender, into the gorgeous homosocial romance of two men on a mesa in New Mexico; the other across sexuality as well as gender, into the conventional but enabling marriage of a historian of the Spanish in the New World. As with the insubordinating parentheses in Proust, the male-homosocial romance represents at the same time the *inside* lining of the heterosexual bond (since the two segments of the domestic story flank their own history in the flashback interpolation of the mesa story) and equally its *exterior* landscape (since the Blue Mesa romance of Tom *Out*land, true to his name, is also figured in the "single square window, swinging outward on hinges"[8] that vents from the Professor's attic study the asphyxiating gas of his stove, and admits to it the "long, blue, hazy smear" (p. 28)—the view of Lake Michigan "like an open door that nobody could shut" (p. 30)—that makes the empowering distance for his intellectual achievements and

7. Leo Bersani, "The Culture of Redemption: Marcel Proust and Melanie Klein," *Critical Inquiry* 12 (Winter 1986): 416. (A slightly revised version of this quotation also appears in Bersani's *The Culture of Redemption,* p. 24.)

8. Willa Cather, *The Professor's House* (New York: Vintage Books, 1973), p. 16. Further citations refer to the same edition.

desires.) The room of one's own, in short, as a room with a view: In this text the crossing of the upstairs/downstairs vertical axis of heterosexual domesticity *by* the space-clearing dash of a male-male romance may some- how refract and decompress the conditions of a lesbian love and creativity.

What become visible in this double refraction are in the first place, of course, the shadows of the brutal suppressions by which a lesbian love did not in Willa Cather's time and culture freely become visible as itself. Still, we can look for affordances offered by that love to these particular refrac- tions. On the one hand, there is the distinctive sensuality attributed to the male homosocial romance, for instance, its extravagant loyalties aerated by extroversion, eye-hunger, and inexpressiveness. On the other there is the canny and manipulative relation to permeable privacies, of a tradition of heterosexual marriage in which marriage is inveterately reconstituted in relation to its Others, in the age-old economy of the Muse represented in this book by the memory of the dead boy Tom Outland. Each of these re- fractions seems moreover to be a way of telling the same story, the story of how expensive and wasteful a thing creative energy is, and how intimately rooted in a plot of betrayal or exploitation: Tom Outland's conscious and empowering betrayal of his beloved friend on the Mesa; the Professor's self-deceived expropriation of the labor, vitality, and money of his wife and daughters. Certainly there is no reason why the right artist at the right moment cannot embody these truths that seem to be, among other things, lesbian truths, directly in the plots of lesbian desire, nurturance, betrayal, exploitation, creativity from which they here seem to have sprung. Then why *not* here? Among the reasons Cather may have had for not doing so must have been, besides the danger to herself and her own enabling privacy, the danger also from this particular steely and un-utopic plot to the early and still-fragile development of any lesbian plot as a public possibility for carrying value and sustaining narrative. She may also, however, have liked the advantage there was to be taken of these other, refracted plots as the aptest carriers, brewed in the acid nuance of centuries, for the exposition precisely of exploitation and betrayal. Anyone who knows how to read anything is experienced in reading, for instance, the story of a husband; we have all read that one all of our lives, however resentfully or maybe all the better because resentfully. Our skills are honed to hear its finest vibrations and hollownesses: the constitutive one-sidedness of the story, the self-pity masquerading as toughness, self-ignorance as clear-sightedness, abject ex- actions as rugged independence. We hear this and yet, in our immemorial heterosexist intimacy with this plot, we are also practiced in how not to stop listening there. We are experienced at looking in this place, even, for unrecognized pockets of value and vitality that can hit out in unpredictable directions.

The Professor's House ends with the exaction of such a reading, I think. Alone in his study while his family has summered in Europe, the Professor now feels he has taken the manful measure of the hard truths of life itself, the ones most alien, he insists, to the venal and feminized values of familial (read female) life that had supported and must return to envelop him.

> He had never learned to live without delight. And he would have to learn to. . . .
>
> . . . He doubted whether his family would ever realize that he was not the same man they had said good-bye to; they would be too happily preoccupied with their own affairs. If his apathy hurt them, they could not possibly be so much hurt as he had been already. At least, he felt the ground under his feet. He thought he knew where he was, and that he could face with fortitude the *Berengaria* and the future. (pp. 282–83)

"Face with fortitude the *Berengaria* and the future," the novel's last words. Conceivably it is the very coarseness and obviousness of the gender asymmetry of heterosexual marriage that licenses, here, a certain sadism of suspicious reading, that forces us not to take this straight as the zero-degree of revelation in Willa Cather's novel. It might impel us, for instance, to want to lay our ear against the tight-stretched drum of that stirring nonsense word, the she-vessel's name *Berengaria*. (A nonsense word insofar as it is a proper noun, it has for the same reason its own attachments: Berengaria was the wife of Richard the Lion-Hearted, who was known for preferring to her intimacy that of men, including, legendarily, his young minstrel.) *Berengaria*—a very mother lode of anagrammatic *energia;* within language, a force of nature, a force of cleavage. Underneath the regimented grammatic f-f-fortitudes of the heterosexist ordering of marriage, there are audible in this alphabet the more purely semantic germs of any vital possibility: *Berengaria,* ship of women: the {green} {aria}, the {eager} {brain}, the {bearing} and the {bairn}, the {raring} {engine}, the {bargain} {binge}, the {ban} and {bar}, the {garbage}, the {barrage} of {anger}, the {bare} {grin}, the {rage} to {err}, the {rare} {grab} for {being}, the {begin} and {rebegin} {again}.

"Willa Cather and Others" was written for a Modern Language Association panel in 1987. The people who sparked my interest in "Paul's Case" were an English teacher, Rita Kosofsky, and a graduate student, Eric Peterson, whose English teacher I was, and on whose work on "Paul's Case" this reading tries to build.

A POEM

IS BEING

WRITTEN

This essay was written late: twenty-seven years late, to the extent that it represents a claim for respectful attention to the intellectual and artistic life of a nine-year-old child, Eve Kosofsky. But it would be fairer to admit (and I can testify to this, since my acquaintance with the person named has been continuous) that her claim for attention to her intellectual and artistic life has in fact, exceptionally, persisted through every day of these twenty-seven years and more, as unremittingly and forcefully as self-respect would permit and very often a good deal more so. What comes late, here, is then not her claim itself, which both deserves and was denied respect *because of its very commonplaceness,* but the rhetorical ground on which alone it can be made audible, which is unfortunately and misleadingly the ground of exception. She is allowed to speak, or I to speak of her, only here in the space of professional success and of hyperconscious virtuosity, conscious not least of the unusually narrow stylistic demands that hedge about any language that treats one's own past. Only, I am aware—I learn this from *Great Expectations,* to name no nearer teachers—only an elegantly ostentatious chastity of style can at all neutralize the fearful (*self*-fearful) and projective squeamishness that for successful adults churns around the seeing displayed of children in their ambition and thought and grievance, in their bodies, in their art, and, indeed, in the simple fact of their oppression. Fortunately, the visibly chastised is by now my favorite style.

The title of this essay, "A Poem Is Being Written," obviously means to associate the shifty passive voice of a famous title of Freud's, "A Child

Is Being Beaten," with the general question of poetry—with the scene of poetry writing, and with the tableau of the poem itself. That association will be at the center of my project, in a discussion that has two parts: the first, a relatively ungendered one, on spanking and poetry; and the second, springing from that, more explicitly and complexly gendered, on the narrative and epistemological misrecognitions that surround *the absence of a discourse*—a discourse, specifically, of female anal eroticism. The title "A Poem Is Being Written" has, however, also, a more specific referent whose existence I should admit here: a particular unfinished long narrative poem, a sulky problem child of my own, itself going on nine, "child" in *many* senses "of my right hand," and joy, whose swollen proportions are only— in a society that hates fat kids—less a shaming badge of my maternal deficits of nurture and discipline than has been its virtual failure in the last two years to grow any *more*. The hideousness of the judgment (on mothers, on children and their rights) expressed in the facile phrase, "a spoiled child," has been especially haunting to me during this time. *Can* a complex, developing consciousness or articulation, full of its own and others' intelligence and desire, be, simply, at some moment, *spoiled* like food? The possibility is so nauseating that it is no wonder the biblical culture that holds it needs children on whom to project, in whom to embody it. Trying to embody this fantasy neither in the child I was nor in the children I don't have nor in the poem I am struggling not to deform or abandon, my meditation on the punishing of children has occurred forward and backward across years, and across the gaps between poetry and theory—and so I'll need to present it here.

Let me start with a few lines from *The Warm Decembers,* my problem child, a half-written novel-length narrative poem with a Victorian setting, and with *its* problem child, its heroine Beatrix, the child of a misalliance between a wealthy father and a mother who is almost a whore, but whom for some reason the father Henry has married—getting himself disinherited in the process, and reduced to the numbing depths of Grub Street journalism: dictating "The Letter from Italy" from a clammy one-room flat in London. What I want you to hear in these lines is more than anything the theater of this family—the space-making, space-marking chiaroscuro of their need and denial, centering on the nude child seen by coal-light in the cold bath.

> The shrimp light breasting the tub and
> one cheekbone, one browbone of the bulky
> girl there, swam onto
> his spotted, felt-shod foot:

all three in the one narrow space set
for the messenger (a sleepy boy
nudging, in the thickest shadow, outward
against the door his plumy, toppling
head) from the *Weekly Review,* come to fetch
The Letter from Italy, which Henry Martin
dictated beside the tub to Clare. . . .

                  Father should be asleep.
Clare and five-year-old Beatrix usually
did their work and their own toilet by
the blearing, then the blinded, coals
in the hours after his "sleeping-draught"
torpedoed him; but not seldom there were
nights, like that night, when after tumblers
and tumblers full, lungs pulling stony through
the lengths of thrown-back torso in the dark,
in the dark finally his eyelids rose
over the trepanning outraged
regard, dull brush of nacre, still awake,
awake even while through the soft palate
unnerved by sleep bleated and bleated
his lower body's breath. Unbelieving, finally
"*I*'ll wash the child," he'd voice.
A few fresh coals woke the resentful fire;
the red-gilt tub, for once, was warmed,
the child among her father's fingers more
and more wooden, deeply chilled (as when
he gave her lessons, taught her, later, how
to calculate, to translate, how to speak):
if at some such moment a messenger
should come, from the *Review* or *Quiz*
or daily *Quidnunc,* for the night's copy
(which might be Characters of Widows, Strange
Laws of Orientals, Beauties of the Lake
Poets, Is Painting Dead?, On Girls
Today and Dangers to Their Health, whatever would
spread thin and make bank of
those earlier travels, glum years at Cambridge,
and now a strange facility,

the child of delirium and cant),
the little family that had come into
legal existence, economically
exploitable, plastic, declassed, only
perhaps in order to legitimate
this fretted potentially female nudity
in the tin tub, was saved, again, saved
for the next day with its own trickle of light;
cold meat; raw mist; hot drink; sleep denied;
with its rainy glaze of the paternal
words, its own vision glued to the
lanky, blotted slippers of felt.
What, in all that time, did Beatrix
understand about her family's
status and history? Nothing was kept
secret from her or revealed to her;
the things she put together were both more
than wanted, and less than at any rate
Henry had feared. It was from Henry's fears
that Beatrix learned things most. To learn,
meaning at once *to be passive* and *to
be feared,* to have a power never, at
the same time, under her control or in
proportion to any detail of herself,
became a transcendent appetite (transcendent,
i.e. insatiable, because of that detour
around the quick unlandmarked spot
marked simply, on the map, Éclaircissement),
unproportioned as the link
of her cold dinner to cold tummy, or as
the snap of her father's hysterical hand
at her ears, to the expensive languid
vocalizations this enforced. Half-ignorant,
makeshift, absorptive, thus, Beatrix
grew up, clumsy to the eye, genteel
to the ear, mysterious to herself, because
to be a woman, what is it if not
to serve the purposes (labor, display,
mobility, verbal recuperation) of a
particular class? Beatrix, almost

lost through the very coarseness of
the family's social weave, to class, almost
also escaped some of the waste of sexual
resolution, almost, except for the
imprint of violence.
(*The Warm Decembers*, chap. 3)

Spanking and Poetry

On a recent visit to my parents' house, I found a poetry anthology they
had given me as a present when I was nine. In the few years after age nine
I'd spent a lot of time with this anthology: I remembered it perfectly, including almost every one of the Joan Walsh Anglund illustrations I'd so
much admired and wanted, myself, to be like: white children in bonnets
and smocks and pinafores, round-cheeked but slender, with no noses and
no mouths. Looking back through it, I was able to see what I had mainly
gotten from the anthology as a poet—it was easy to find, since I had gotten it all, as far as I can tell, from a single poem. As far as I can guess now, I
must have spent hours—probably, really, hundreds of hours—reading and
rereading one bad poem by Louis Untermeyer, the editor of the anthology,
a poem called "Disenchanted." I *know* this poem, though oddly, I don't
know it by heart. I know its two-beat line as well as I know my own pulse. I
know it because it's the beat of almost every poem of any ambition I wrote
for ten years after the age of nine. The poem starts,

> Here in the German
>   Fairy forest,
> And here I turn in,
>   I, the poorest
> Son of an aging
>   Humble widow.
> The night is fading;
>   Every shadow
> Conceals a kobold,
>   A gnome's dark eye,
> Or even some troubled
>   Loreley.[1]

1. Louis Untermeyer, ed., *The Golden Treasury of Poetry* (New York: Golden Press, 1959),
p. 168.

And the rest of the poem goes on just as footlingly. But I was genuinely in love with something in this poem: it gave me power, a kind of power I still feel, though I no longer feel it in this poem. The name of that power—I know it now, and I knew it not long after I got this anthology at age nine—is, enjambment.[2]

Or maybe we should go back a bit. When I was a little child the two most rhythmic things that happened to me were spanking and poetry. Not that I got spanked much: my big sister and I and our little brother were good and well-loved children and were not abused children, and it matters that this narrative is about an attentive, emotionally and intellectually generous matrix of nurturance and pedagogy. Still at times certainly, at the most memorable moments of our childlives (except that we barely remember them), suddenly within the quiet and agreeable space of the Biedermeier family culture of upwardly mobile assimilated American Jews in the 1950s there would constitute itself another, breath-holding space, a small temporary visible and glamorizing theater around the immobilized and involuntarily displayed lower body of a child. Of one of us. Not only shame itself, the forbidden, compulsory, and now too late longing to be excused from the eyes of others, which—or the forestalling or denial of which—was and maybe still is the mainspring of my own most characteristic motivations, was centered in this movable theater,[3] but other plummier things as well. Wasn't it after all only the fat, innocent, and, at least in me, rankly profuse *exhibitionism* of an unpunished child that could lend this being compulsorily exhibited its peculiarly rich and lasting fetor of mortification? A primal hunger to be seen was certainly not undone in these punitive moments, but only made inseparable from the paralysis of my own rage and the potency and bland denial of my parents' rage; from the tensely not

2. Enjambment: "a technical term in verse, signifying the carrying on the sense of a line or couplet into the next"; Margaret Drabble, ed., *The Oxford Companion to English Literature*, 5th ed. (New York: Oxford University Press, 1985), p. 320.

3. A lot of the resonance attaching to the notion of "shame" in this essay and some of that attaching to "misrecognition" as well comes from Stanley Cavell's essay, "The Avoidance of Love: A Reading of *King Lear*," in *Must We Mean What We Say?* (New York: Scribners, 1969), pp. 267–353. The interest in theatricality owes something to Michael Fried, *Absorption and Theatricality: Painting and Beholder in the Age of Diderot* (Berkeley: University of California Press, 1980). Robert Stoller gives an extended reading of a sadomasochistic sexual fantasy involving theatrical tableau in *Sexual Excitement* (New York: Pantheon Books, 1979); that book itself, like the therapy it records, would repay analytic treatment as an example of the countertransferences, furies, and political tendentiousnesses that can characterize male feminist uses of psychoanalysis, but among its achievements are a vivid and thematically rich evocation of a cluster of issues similar to those I will be dealing with here.

uncontrolled, repressed and repressive (and yet how speaking) rhythm of blows, or beats; from the tableau itself.[4]

The tableau: I'm struck, as they say, by the importance of French in this story. It's a problem that I sometimes don't know when I learned particular words. But while I know we were always spanked, in a careful orchestration of spontaneity and pageantry, "simply" over the parental lap, the spanking in my imagination (I can only barely stop myself from saying, the spanking that *is* my imagination) has always occurred over a table, a table scaled precisely to the trunk of the child, framing it with a closeness and immobilizing exactitude that defined, for me, both the English word *truncate* and the French, of course, *tableau.* The glamorized, inbreathing theatrical space of the spanking thus contracted to the framing of a single, striped, and sectioned mid-body that wanted to move and mustn't. Did this imaginative contraction happen once or over and over again? Of course I don't know. (I planned for each poem in a booklet I made at eight to be closely framed by a golden proscenium hung with curtains, carefully labeled "The Magic Window.") But I see now—this is in Freud, too—that part of the effect of this concentration of framing is to eject from the *tableau* or table itself, along with every figure but the figure of part of the child, the entire visible mechanism of the gaze to which the child is exposed, the graphic multicharacter drama of infliction and onlooking, the visibly rendered plural possibilities of sadism, voyeurism, horror, *Schadenfreude,* disgust, or even compassion. The redrawing of the frame in such a way as to banish the other characters of the drama after all however de-differentiates and reunites them, does it not, on the hither side of the viewer's or imaginer's own gaze: the decontextualized, legless, and often headless figure of di/ s/play creates in turn a free switchpoint for the identities of subject, object, onlooker, desirer, looker-away. For, at the most schematic, active and passive, but also for roles that far exceed those in resonance—the reactive, for instance, and the impassive.[5]

4. On this, see Alice Miller, *For Your Own Good: Hidden Cruelty in Child-Rearing and the Roots of Violence,* trans. Hildegarde Hannum and Hunter Hannum (New York: Farrar Straus and Giroux, 1983); and *Thou Shalt Not Be Aware: Psychoanalysis and Society's Betrayal of the Child,* trans. Hildegarde Hannum and Hunter Hannum (New York: Farrar Straus and Giroux, 1984).

5. Sigmund Freud, "A Child Is Being Beaten," *Collected Papers of Sigmund Freud,* authorized trans. under the supervision of Joan Rivière, 5 vols. (New York: Basic Books, 1959), 2:172–201. A suggestive discussion of how these roles are acted out in the pageantry of, this time, the operating theater, is to be found in Michael Fried's "Realism, Writing, and Disfiguration in Thomas Eakins's *Gross Clinic,*" *Representations* 9 (Winter 1985): 33–104.

It is in this way, I think, that both the cropped immobilized space of the lyric and the dilated space around it of narrative poetry were constructed in and by me, as well as around and *on* me, through the barely ritualized violence against children that my parents' culture and mine enforced and enforces. The lyric poem, known to the child as such by its beat and by a principle of severe economy (the exactitude with which the frame held the figure)—the lyric poem was both the spanked body, my own body or another one like it for me to watch or punish, and at the same time the very spanking, the rhythmic hand whether hard or subtle of authority itself. What child wouldn't be ravenous for dominion in this place? Among the powers to be won was the power to brazen, to conceal, to savage, to adorn, or to abstract the body of one's own humiliation; or perhaps most wonderful, to *identify with* it, creating with painful love and care, but in a temporality miraculously compressed by the elegancies of language, the distance across which this body in punishment could be endowed with an aura of meaning and attraction—across which, in short, the *compelled* body could be *chosen.* Literally the only theft I remember ever making from another person was of a big green rhyming dictionary (I still use it), from my grandfather, who had used it to write in the autograph book I had at nine,

> With face so cute
> And eyes so bright,
> A nose so pert,
> You're a lovely sight.

What I told myself at the time I felt was simply envy at the virtuosity of this *blason* of my person: I wanted the book that would let me do it, too. What I see I really did with the rhyming dictionary, though, was a bit more radical as a framing of the compelled and immobilized body. The next poem I have from the period after the theft is called, cheerily,

> *Stillborn Child*
>
> Crumpled face
> On a sea
> Of pillowcase.
>
> Greyish skin
> Withered not
> By age or sin,

Creamy smooth.
Expression—
Angel-aloof.

Burning-red
Hair that feels
Singed and dead.

Eyes that see
Nothing. Void
Of expression,

They are faced . . .
(ca. 1961)

You will have noticed the magic cropping of the *tableau* that turns what
might have been the angry self-pity of an eleven-year-old redhead into
something more concentrated and, to me then, adult-sounding—into genuine morbidity. You might also have picked up on the unmistakable presence, in the Kosofsky poem, of the two-beat Untermeyer rhythm with
which I have said this writer was by this time completely infused. The thing
in this embarrassing poem that to me now spells a certain resistance and
heroism, though, the real place of challenge both to the static *blason* of
female beauty and to the petrifying *tableau* of punishment, is the new presence in the poem, also from Untermeyer, of enjambment.

All this childish French. But I was starting French in school, at eleven,
with the ravishing Monsieur O., so I knew about parts of the body, *la
jambe,* and foods, *le jambon*—and because I loved French (because I was
eleven and ambitious but not least because I loved the "darkly handsome"
and ingenious Monsieur O. with his ramrod bearing), I knew *enjambment,*
not just for a technical word in the introduction to my rhyming dictionary,
but for a physical gesture of the limbs, of the flanks, the ham. I thought
then, too—in fact I thought it until I checked my dictionary just today—
that a *doorjamb,* for instance, was the thing one wedged in the door to keep
it open, a doorstop. From all this I visualized *enjambment* very clearly as
not only (what my French dictionary now tells me) the poetic gesture of
straddling lines *together* syntactically, but also a pushing *apart* of lines. In
terms of the beat(ing) of the poem, enjambment was, in this fantasy that
shaped my poetic, the thrusting up out of the picture plane in protest by
the poem's body of a syntactic thigh or shank that would intercept, would
retard the numbered blows: would momentarily wedge apart with sense

the hammering iteration of rhythm. Would say no. The physical and moral splendor of this instinctive gesture, to my imagination both then and, I must say, now, depends of course on the assumed framework of—what I utterly was—an almost absolutely ruly and obedient child.[6]

Still, if I had, over a period of a decade or more, a kind of craze for the heroics of enjambment (and I can find poems where simply nothing else seems to be going on), that must have functioned as a sort of flag of bravada resistance under which a far more complex and recuperative negotiation was in process. Even enjambment itself, after all, while offering the to me thrilling image of gut- or groin-level resistance or impedance, worked, in actual poems, not through the Kristevan semiotic but quite the opposite, through whatever was most abstract and cognitively under control in the poem, through the forward-looking and distributive pressure of the syntactic. With that abstraction I learned to identify. The effect I fell in love with enjambment for creating was one of decisiveness. But in the grammar of the close-cropped lyric tableau, shifty *because* simplified, *because* singular, decision became a power not to resist the poem but, more simply, to decide *it;* a power, if you like, to spank. A trick of discursive authority, grotesque in a preteen or a teenager, became familiar to me, became my utterance in prose as in poetry: and if I have in some ways grown into it since then, what I have never begun to be able to imagine is a way to shed or circumvent it. The main component of this "authority" seems to be the

6. There are two more things to point out about this image. One is a pictorial reminder: that the interposed limb or ham is a truncated, in fact an amputated one. Another is the specific grounding of this physical gesture, not only in the child culture of the spanking, but specifically in the girl culture (of that class and time) of ballet. In many ways ballet itself functioned for many girls, including me, very much as poetry did for me, as another archmediator of one's relation to half-ritualized violence: like poetry, it was a rhythmic, prestigious, exhibitionistic, and highly theatricalized way of choosing the compelled and displayed body—here, an intensively though impossibly gendered one. (The egalitarian bliss of girls undressing together, my nicest memory of "ballet," somehow turned through this culture into the rapt recital and celebration of a rigorously meritocratic hierarchy—corps de ballet, ballerina, prima ballerina, prima ballerina *assoluta*—that was the only plot, aside from heterosexual love, of the ballet books we gobbled up in series.) If I can't remember any more the name of the particular kind of *battement* that's being performed by the prone figure in my tableau of enjambment, I can still remember the founding moment in a ballet class of my articulable sense of the transfigurative (i.e., misrecognition-creating) potential of any art, a dictum only the more gravely received for being directed, not to my own, but to my big sister's class, which I was visiting at age five or so: "The ballerina's limbs *must* look vulnerable, but to do this they *must* be strong as iron." I learned early, also, that ballerinas are nearly always in pain and mustn't ever show it. The potential for sustained and productive, if costly, play between power and impotence, through the medium of sanctioned spectacle, was not lost on me or on many other girls.

presentation of a piece of writing as the now-under-control palimpsest of some earlier, plurivocal drama or struggle: among tones, among dictions, among genres. The visible marks of solicitous care and of self-repression, the scrupulously almost not legible map of exorbitance half-erased by discipline, the very "careful [one might add very pleasurable] orchestration of spontaneity and pageantry" (the same with which one's parents took one over their lap): these stigmata of "decisiveness" in and authority over one's language are recognizable as such by their family resemblance to the power, rage, and assault that parents present to the child with a demand for compulsory misrecognition of them as discretion and love.[7]

Our story so far, insofar as it *is* a story, seems to be about a generic, i.e., ungendered, "child" that finds its way, through a drama of the enactment and occlusion of power and rage, to a shiftily identificatory relation to, generically, the genre "lyric poem." But the trajectory of this story must, as I at least am aware, finally be toward an adult of differentiated gender whose poetic energies seem inextricable from a far different genre, the long narrative. It is not, either, one story—there are more and less direct, more and less inclusive, simply different trajectories from there to here; and if it is easiest to narrate, now, through the sexual (because sexuality—as, necessarily, retardation; as, necessarily, perversion—is the right vehicle for these crossings of part against whole, of incipience against retroscission) then the sexual, too, must be various even at a given site. A pornographic paperback I once got, subclass subliterate pseudo-scientific, liked to classify its characters' various activities among the "perversia," and perhaps we could proceed under worse aegises than the prurient inventiveness of that plural.

Let me, then, jump ahead here for a moment along the simplest of the trajectories to hand, the directly sadomasochistic one whose perversion lies in its exact, its *retardataire* simplicity. Not juvenilia at all but the charged-up work of a twenty-four-year-old graduate student, the poem that can represent it is the one that enacts in the most literal terms the dilation of view, the generic leap from the tightly framed tableau of parental punishment to the social and institutional framing as narrative of exactly the same scene, with a heartbeat percussiveness toward all "the visibly rendered" and newly excited "plural possibilities of sadism, voyeurism, horror, *Schadenfreude*, disgust, or even compassion." Far—more than a decade, as we'll see—from being a first narrative poem, "Lost Letter" was nonetheless the most significant manifesto for its writer in the so far unretraceable direction of

---

7. For this reason, I always find it hard to figure the polylogic, at least in my writing, as liberatory *as opposed to* disciplinary.

the ever-longer narrative, and the increasingly dialogic. That its intensely self-reflexive ambitions included the directly pornographic (and in the first place, autoerotic: the unwritten essay title floating around in my head at that time was "The Hair on the Palm at the End of the Mind") can make me, in some ways, queasier now than it did in 1974, but "Lost Letter" was never the most possible poem whether to classify, to show, to publish, to secret, to address, to throw away entirely.[8] It began in low key, a letter to a former professor from an apprentice one:

Lost Letter

I.

My letter cools its heels in a suburb
of Paris—there's a mail strike now—
or is lost in some confusion. And rightly:

8. An autobiographical project that emphasizes discursive formations around sexuality makes one aware of startlingly rapid shifts in the thinkable and sayable. "Lost Letter," for a notable example, first written on a grounding of (in me) the most rudimentary feminist consciousness, is read and reproduced now by me through a far fuller one, including the ambient though considerably resisted influence of an antipornography movement whose defining instance of "pornography" is pretty much exactly the sadomasochistic, female-punishing tableau enacted in the poem. However defiant its reproduction here may feel in such a cultural context, the possible reading of it is now nonetheless heavily marked by those emphases. Other changes have intervened perhaps less articulably but perhaps also as powerfully, such as the audible eddies of decolonializing imperialism and nationalism: the international level of dependence on torture has risen, and the representational machines for publicizing its means and effects are busier and more successful than a decade ago, with complicated results of imagery, ideology, passion, fear, resistance. None of this, naturally, makes me feel more simply complacent about "Lost Letter." But (and this is not the only moment of this essay where this assertion may need to be made): if we know from Foucault among others that the very energy and ingenuity of the radical drive toward what feels like *truthtelling*, in a given age, are likely to contribute to projects of enforcement as much as to those of transformation, that doesn't on the other hand make *attenuation* any the less necessarily a repressive act. Thus rather than present the writer in this essay as an exemplary figure on whom certain airbrushing practices would have had to be used, it's seemed more productive to preserve and to invite in the reader a distance from her that licensed freely the "operations" of knowledge, even a certain examinative sadism, such as I would have found even harder with any other woman writer. At the same time (isn't it the telltale thing about pornographers that they can't stop moralizing?) one might as well make instructively explicit, as one can hardly render inoperative, the truth that one pays for the energy one's "perversia" lend to her writing by certain moments of giving away with both hands, at the textual places where these energies break the surface, the poise or secret or charade of authorial control over the identifications of "the" reader: the scandal of any sexuality (that it is so wastefully particular) and the particular scandal of a sadomasochistic one (that it so desublimates the enabling conditions of any sexuality) coming together with a projectile force which it would be reckless to make much pretense of keeping under patronage.

it was a confused letter anyhow.
I thanked you (in it) for a recommendation
(you've been writing them since I was a freshman).

"I'm teaching writing now," I said, "isn't it
uncanny the authoritativeness
of puppy genius, and so distant."

No wonder we're uncomfortable, I meant:
the pain of teaching being so akin
the pain of studenting, of envy and arousal

at language barely meant for our proper eyes and ears.
Both shy, we nowadays wolf down
our letters as greedily as if

2.

they might shift everything. And they might, but not toward us.
The time I spend writing lost letters!
Weeks, sometimes, for yours, or witness

in pages of minute, pornographic script
a novel in letters, about a man
hurrying to the dentist, doctor, or, it may be, shrink

past an open door in a downtown medical building
who glimpses in a waiting room a naked girl
submitting to something evidently

jazzy and frightening in the way of punishment.
He's appalled, he can't watch it,
but he's recognized her, a student he's fond of,

and, once home, writes her a hesitant note.
"If you ever feel it would help to talk
please consider me at your disposal;

if, though, I've completely misunderstood
the fleeting scene, forgive me, Eve, it just
seemed you might need something as obvious as this:

Nobody, human, would turn from you for your
having suffered some sickening discipline. . . .
how long do you expect it to continue?"

The poem asks for itself the question anyone would ask:

What kind of novelistic world is this
where college women, on their own,
find their way to offices in downtown medical buildings

for jazzy, frightening punishment? What crimes
do college women know how to commit,
and where, in a college town, is such an authority

as could force the firm, unspeakably reluctant feet
up sloping flights of dingy stairs, and in?
What keeps the face, under punishment, impassive,

the nude body motionless as ordered
though trembling? Not, for sure, the individual volition,
nor yet the school. A shadowy arm

maybe of Parenthood, mysterious and known to all,
ramifying in every city
with conspiratory potency. Of course

in the real pornography these questions are
—*comment dit-on*—extratextual, moot;
mute, taken obscenely for granted, or part of the *frisson:*

so much so, the mention of them even here is recklessness.
The compulsion in the story's real;
"There is real violence being done right now,"

4.
he writes back later, "and it's not to my sensibilities.
Please don't write nonsense, no outside infliction on you,
I don't care how messy it is (for it's messiness you're talking about,

reading about someone else's pain isn't nearly
as assaultive as reading about their mess) could even threaten
my sympathy and identification with you now.

I don't much care about being told these things.
Or more accurately, my reaction to being told
is just what you might guess, in a more lit-critical mood:

i.e., a not unfamiliar mix of pain and anger
and disgust, yes, and furtive arousal, also yes, but so
what? Shit, Eve, we can't

let our friends withdraw from us, or ourselves withdraw from them,
just because we've complicated reasons
for caring about each other." I'm embarrassed too

at the wishfulness of this, but please remember
the wash of helplessness that's bringing them together,
everything outside the letters spelling shame and terror.

The mutual letters, whereon, embark on *their* spelling of shame and terror,
some of which I can skip here: the repetitious pornography of the continuing *tableau* itself and that of its insistent "empathetic" reproduction in the
cool, sappy, imitable voice of the teacher, twining into ever more prurient
fine tunings about his oblique involvement—

    "I'm not sure how to feel

about having precipitated a crisis in the room
by appearing there at the doorway shattering
your enforced composure—aside from sorry:

it's almost a relief, though this is being selfish,
to find my part in the savaging was that direct,
acknowledged by you at the time, not just a voyeuristic

complicity. . . .
I guess I thought you hadn't seen me, I guess
I thought the rigid, sightless gaze was your defense

against the nakedness below, the pain to come: I didn't guess
you were in pain already, there, and struggling"

—and quarrels over the affective evacuation in his letters of the language of
assault (much requoted, on that pretext, back and forth). "But," she protests,

"that's me—that naked trunk that's bent and tied
over the abdomen-high table, waiting for stripes;
    . . . That's me, all these weary nights

waking to my own screams five and six times before dawn
afraid to fall asleep, afraid to wake and find
a note of summons slipped under the door

to climb the steps again in sober winter daylight
arriving, shaking, at the right floor, for more. . . ."

. . . But later, sorry and frightened,
"Please don't think I'm angry at you, in fact

I don't know what I'd do without you. That's literally true.
My sense of the world is broken and past fixing.
Don't think I'm angry, either, that you saw me shamed—

the truth is yesterday, under the whip,
it came to me in a thud of longing—or nostalgia—
how much I'd give to see once more that my

naked parts could still embarrass and appall. . . .
　　As for talking instead of writing, I'm not sure,
but I mean to be in class tomorrow so we'll see.

Please, by the way, don't call on me in class.
I'll volunteer if I can talk, but may be struggling
just to sit still. I'll see you soon.

7.
P.S. Forgive the way I began the letter."
　The next letter, from him, next afternoon:
"I meant to stop you after class but it was clear

it wasn't going well for you. Every time
one of the boys in the class moved or spoke you flinched
and went white. It was awful. I guess

it hadn't been so present to me, before,
how much you've been abused just lately at the hands
of men specifically. It makes me look down at my own"

which he really does, long hairy ones, "terrified.
You don't seem to mind much when I write,
but I was afraid you'd wince away from me too

and you shouldn't have to do that, so let's write.
If you mean to stay in school though you'll have to come to class.
And for your good, you need to work

on overcoming the horror of men. And I don't think
I say that just because I'm a man, certainly not
from not knowing what horrors it's a response to . . .

but when I write it turns into 'a whipping'

and you think I'm being casual or debonair
at your expense. But no, I'm not.
Your letter can bring fresh shocks of impotence,

urgency, vicarious humiliation, but even
without a word from you further, I remain yours truly:
the fixed slave of your continuing punishment."

8.
It's hard overcoming the horror of men
when almost daily, for a long time, she'll have to seek out
and penetrate the blank ugly downtown building

to discover subtle and blindingly new
accesses to her of pain, of dread, of
weariness, new forms of nakedness,

and old impassivity fresh and fresh imposed.
They never—the pair of them—learn
to talk it over, so the shrewd

punishing repetitious letters shuttle back and forth
with all the comfort there is;
the shy obsession grows, but never turns,

for them, toward bed. Besides
every orifice in her is so fatigued!
Just twice more, headed for his therapist,

the man takes the slow stairs past the open door,
the first time seeing a boy in a corset
and with stripes, being baited to impermissible tears.

That night the man himself is waked by his own screams
but smothers them to lie in silent tears
trying not to wake his wife with the sheer terror.

The second time his dreadful expectation is rebuked
for there she is. This time he isn't noticed
since the girl is positioned to front the wall, away from the door

her face not being, this time, the focus of attention.
He never tells her what he's seen
nor the therapist, though the therapist is observant

of a particular stiff tenderness as the man sits down
and some semaphore in the brow, and is savvy enough
to be grateful there are things he genuinely can't interpret.

9.
Like the detective's gift, or shrink's, the poet's gift
is blank fatuity and no hint of anger.
As if this resourceful immobility

could, for more than the riveted instant, assuage
the storms of anger that travel around poets!
I've been reading an anthology of recent poets

so as to sound, in conversation, like a poet
so as to get a job teaching young poets
the scopophilic and exhibitionistic transports

that are—no kidding—what make it beautiful.
I haven't found any other poems like this one,
neither as risky nor as unnecessary.

In fact the fierceness of my love for these pages,
all these pages, is the least oblique thing going, here.
I haven't overheard them or hurried past them

or hidden my hand from them for fear
of their flinching from it. Nothing's without
obliquity, pain itself is not, language

about pain least of all, but the shame itself
of privacy should give place with a thud
of longing to this much, this good, attention.
(1974)

    The more than exhibitionism, the blissful new vocational pride of this
poem (much though I *have,* in this reproduction of it, hurried past); the
wracking and politically expensive divorce of the tightly zoned and abjected female body from the enfranchised female voice; the many filaments
with which the poem binds its personnel to their multiple and changing
stations in the culture of the "examination";[9] the equivocal address of this

9. In autobiographic retrospect, the particular manifestations of this that crowd most urgently
are a bad "oral" exam the year before, and, in a somewhat longer perspective, the intensive
delegation of enforcing power to the medical profession—their *point d'appui,* the theater of

letter by which the demarcation of private from public is erotically exploited even as it is generically defied or transformed; the po-faced relish
of mimicry that confuses the power relations of observer and observed; the
half-occluded aggressiveness with which the autoerotic enactment of the
poem both provokes and deprecates the fury of the impinged-on reader,
and the fatuous insistence with which it solicits and pretends to ignore the
desire that is to be the unnamed engine of that fury—all are condensed
in the "thud / of longing" with which the contracted, bodily siting of the
drama of enjambment is displaced back outward onto the sheepish hungry
gaze from the wedged-open door.

Let me, however, retreat from that simplified trajectory, the adult annunciation, to leave this section with two girls at a threshold yet uncrossed.
One is Eve Kosofsky, a week short of age twelve in 1962. The other, in *The
Warm Decembers,* is Beatrix, by now a bulky teenager, whose father has
recently died: she is running away from her mother's boardinghouse at the
seashore, back to London.

> It was August.
>   Father, she thought.
> The dome of St. Paul's spread silvery
> over her inward eye to welcome her;
> under her feet the soft earth breathed shortly,
> as she did. Short and violent bits of languages
> she'd learned of him for his plagiarisms, spoke
> in her, "Geboren in der Flucht," I almost
> stifled in the smoke, "Des Vaters höchste
> Furcht die an das Licht gedrungen,"
> and while she could smell the ocean she tore
> to the west, the moon behind her
> inking the sandy road ahead of her
> and voyaging even faster and more
> silently there. It suited her, to be
> nothing in the negligent
> and devious light, meeting no one, not
> human, silent—her father
> would have said, she imagined, like a horse.

---

the waiting and examining rooms—over the punitive gendered discourse of fat (and see note
15, below). No one who's been to a gynecologist, either, should feel like a stranger in the
waiting room of this poem.

One on San Marco—gilded, scored.
Dull traces of the harness.
Her haunches brushed with mercury.
(*The Warm Decembers*, chap. 3)

Fundamental Misrecognitions

My fella said to me, Sophie, the
problem with you is, you've got no
tits, plus your pussy's too tight.
I told him Harry, get off my back.
—Sophie Tucker[10]

If the second part of my story, a more tortuously inclusive trajectory toward the narrative poem, the gendered and sexed adult, cannot be read as directly as the first off the tableau of the small, bounded, and violence-inscribed torso, that fact might help us be suspicious about the costs of the too-early contraction of view to that torso. Isn't anger almost necessarily the most diachronic, the most narrative of emotions, the one most *necessarily* mistranslated in the freeze-frame? Still, I know nothing more promising to do with the obsessional image of paralysis than the necessarily self-ignorant task of reading it—cheered by the speculation of a brother writer, about the sensation novel, that it may be permitted "to 'say' certain things for which our culture . . . has yet to develop another language" by exactly "the non-recognition that thus obtains between our sensations and their narrative thematization."[11]

The Palimpsest

You must have dropped off on top of the covers
after only a couple of brief throes, for when I
found you in the morning you were just treading

up toward consciousness—still in the costume of your obsession.
Under a latex body suit—rubber corsets.
Then chalk and crayon hieroglyphs to the flesh.

These showed, on the front of the thigh, marks of claws

10. Probably apocryphal; attributed to Tucker by Bette Midler, in her nightclub act.
11. D. A. Miller, *The Novel and the Police* (Berkeley: University of California Press, 1988), pp. 146–91.

and bits of leopard fur; and at the back
uneven horizontal stripes in red.

How different if we had found you in a whipping orgy for real.
How different if some biography had told as a true fact
how puckered, long-healed, horizontal scars

were found in that place by the impassive coroner—
the biographer mentioning, what were really unspeakable,
some corresponding documents: your life in the family, your school.

In fact the scars don't answer to the wounds.
You surface, hence, in the narrow warmth of meaning,
being peeled and washed on the coverlet where found.
(1974)

I can see, at this point, three different non- or misrecognitions condensed
in the lyric tableau of the beaten child—or in my "identification with" it.
The first is the one I sketched above, more fully discussed by Alice Miller in
For Your Own Good and *Thou Shalt Not Be Aware*, the learned misrecognition of injury as nurture, and the subsumptive working through that of
the maturing child toward a position of sadism and cultural authority.[12] (As

---

12. I use a closely related narrative of misrecognition—but one tied to Ferenczi and the
controversy around the seduction theory, referring to "sexuality" and seduction in place of
"sadism" and punishment—in a novella-length poem, "Trace at 46" (1977):
    A child "loves" an adult, who grows confused
by the child's advances, and loves it back,
screwing it, then panicking—repeatedly. The child
        can't recognize
as forms of love either the blind drive toward
the genitals, or the panicky silence: but swamped
by the adult's wild need—by its own anxiety—by
a kind of off-rhyme between the new sensations
        that hurt
and old, climaxless sensations it knows it likes—
helpless to turn away the new love, helpless to
accept it, helpless to keep the adult
from panicky silences or sudden inattention,
the child finally invents from its own baffled
heart a new, expensive theatrics:
a hallucinatory mimicry
of the adult, by which within the child
a second adult is made; whose function

a child, I hated and envied the frequent and apparently dégagé use my parents liked to make of the word *humiliating,* a word that seemed so pivotal to my life that I could not believe it could *not* be to theirs. Now I enjoy—and *enjoy* is the verb I mean—the shared right to act and speak as if I, too, hear and use it as just another word, and to pretend to imagine that others, including children, can do so as well.)

A thematics of, not only the violence-marked body, but specifically the *legible* body, which I have alluded to but not really developed here, has additionally its own paths of nonrecognition and expropriability. These involve among other things the directionality of the body: not in this case the frontality (to the viewer) of the *tableau* but the backness (from the child) of the backside. If I am, supposedly, alone of my species in actually remembering my first reading of *David Copperfield,* that is because the reading was itself truncated—cut off, then *put* off for a couple of decades—by the intensity of my pain, or excitement, or something, at the image of a child made to wear *on his back* a placard, *"Take care of him. He bites."* [13] There is always

---

being to put under the child's control what is
    strangest
in the adult and most at war with itself or the child
—things every adult has, like remorse, harshness,
stupor, intellectual longing, fury—as well as the
    signs
of genital desire—the child pours all its strength
into the animation of a perfect inner stranger,
    only swerving
it a little, always, toward an implausible childish
tenderness from which every encounter
with the real unchanged adult saps the pretense
of mimetic force. The child
whose expedient this is, changes, forgetting itself.
From a mortal trance which it wishes
always to deepen, the small anxious automaton throws
different, intermittent voices—different tongues—
even silences that aren't its own—everything that
    denotes
assent, but whose structure is resistance, to
the love a child can't use.

Diacritics 10 (March 1980): 11–12; a lot of the language here is borrowed from Sandor Ferenczi, "Confusion of Tongues Between Adults and the Child: The Language of Tenderness and of Passion."

13. Charles Dickens, *The Personal History of David Copperfield,* ed. Trevor Blount (Harmondsworth, Middlesex: Penguin Books, 1966), pp. 130–31. My reading of this moment in Dickens comes from D. A. Miller, *The Novel and the Police,* pp. 192–220; more generally, a

a potential for a terrifying involuntarity of meaning, in the body of a child. The adult judgment that children should be seen and not heard—should, like the lyric poem, not mean (their own meaning) but be (the beheld meaning)—is a way adults have of grappling for a minimal control over this, as is the apparent popularity of inscribing children as a form of punishment or initiation.[14] The aptitude of the child's body to represent, among other things, the fears, furies, appetites, and losses of the people around it, back to themselves and out to others, is terrifying perhaps in the first place to them, but with a terror the child itself learns with great ease and anyway with a lot of help. And this leakage or involuntarity of meaning, the seat of this form of vulnerability, is easily located in *the behind*. One has after all behind one, by this time, something significant: the place that is signally not under one's own ocular control and also the site, by "no accident," of the memorializing outerworks of an earliest struggle, bowel training, over private excitations, adopted controls, the uses of shame, and the rhythms of productivity.

The two-sidedness of the insulted body, too, once it has come to represent the uncontrollable *meaningfulness* of the body, becomes as well a more than latent image of its necessary *temporality,* of change as well as vicariousness in the place of the subject. It is one place to begin with narrative. As the placarded David worries: "Whether it was possible for people to see me or not, I always fancied that somebody was reading it. It was no relief to turn around and find nobody; for *wherever my back was, there I imagined somebody always to be.*" Erotically, too, here, the potential excitations of something or somebody endlessly at, on, in one's back, when they become palpable at all, must do so not only along with but *through* an unappeasable anxiety (or alternatively, depression) and shame of visibility, by which one is eyelessly inducted as well into the "productive" drives and rhythms appropriate to (appropriable by) a particular economic system.[15]

---

lot of the motives and moves of the present essay, including some language from chapter 8 of *The Warm Decembers* reproduced below, are responsive to that essay.

14. See, for instance, Rudyard Kipling, "Baa Baa, Black Sheep," in *The Writings in Prose and Verse of Rudyard Kipling,* 35 vols. (New York, 1907), 6:323–68; or Charlotte Brontë, *Jane Eyre,* 2 vols. (Edinburgh: John Gray, 1905), vol. 1, p. 118 (chap. 8).

15. Cf. my *Between Men: English Literature and Male Homosocial Desire* (New York: Columbia University Press, 1985), esp. chaps. 6, 9, 10. I will be discussing infra a major-key, nonvisual way in which this misrecognition-from-the-back also represents an induction into a particular *gender* system. But it should be said here as well that no dynamic of visibility can, in our society, itself fail to be heavily gendered. A relevant mechanism for this has to do with the (very intensively gendered in our society) "aesthetics" and "medicine," i.e., prescriptive ethics, of fat. A compulsive joke of my (extremely slender) mother's was to murmur to us, at the sight of any fat woman walking away, "Is that how I look in shorts?" The near

From later in *The Warm Decembers,* a chapter begins with this story of violence to an ungendered child, a kind of "floating" story that attaches itself differentially to me, to Beatrix, to a couple of other characters whom you'll see . . .

> To have been depressed in early life.
> Imagine a child wetting its bed.
> And say the family's poor, the beds are shared,
> the washing's done in buckets and by hand,
> the drying sheet smothers the attic room.
>
> Whose crazy father then decides:
> This is a child who "must not" be given water.
> Or: it is dangerous to let this child sleep.
> The awful logic nods only when he does—
> and then the parching child nods off
> in sleep that's only waking, waking.
> Waking to violence or the expensive wet
> that makes violence. And say the child survives
> and finds, somewhere, an art.
>
> Waking in the morning, I remember first
> I'm grown up. I have some money and a car
> and anything I want, to cook and eat,
> and (in the horrid, doggerel blank verse
> in which I—no, not "think"—but breathe, and represent

---

impossibility for the adult European female body of maintaining the twentieth-century disciplinary ideal of constant total muscular control, over the ambulatory rear end, makes it especially "meaningful" as the locus—forever hidden from one's own eyes—of an always potentially discrediting scandal, the scandal of the very materiality and difference of female bodies. I understand that there are strong cultural differences in these perceptions: for instance that African American eyes find it easier to see women's bodies—and through them, women's character, their sway, their sexuality, these not being figured either as minimalness or as tight muscular control—as being strengthened, not discredited by their substance. (Still, "the sexual politics of the ass are not identical to the sexual politics of the asshole" [see infra]: I am told, though I can't make a generalization of my own about this, that a high valuation in African American culture of women's substance including the rear end coexists with an attitude toward anal eroticism that is as severe as the Euro-American. "You've got the right church," Bessie Smith sings, "but the wrong pew.") But for most Euro-American women, and certainly for middle-class ones, the inextricability of a dynamic of shame from every issue of body size, i.e., substance and *visibility* (and this not only, although of course markedly, for fat women) would be hard to overstate. And a dynamic of shame means, perhaps necessarily, a narrative about what lives and senses and "speaks" behind one's back.

continually to my own ear the place
of my unthinkingness) repeats, repeats
some vapid version of a Shakespeare phrase,
"Yet Edmund was beloved."
Waking alone, yet E—— is beloved.
Also this: "an important writer of
fiction and poetry,—"
                              of *criticism*
and poetry, of course it's meant to say,
but "fiction," in this empty register,
scans, so "fiction" in my head it always is. . . .

Waking as an adult, now, who has an art.

An adult, I mean, who's not depressed.
For whom a vacant, distended, paper-light globe
called "gratitude," fills up the inner space
(gratitude as it were for water and for sleep:
for being able not to loathe "the sweet approach
of even or morn, or sight of vernal bloom,"
"or flocks, or herds")
—gratitude without an object, too, since these
good things—love of our life—
are our own true birthright if we've *any* right.
Gratitude, *positive happiness,* not the less for that.
Positive, meaning not good necessarily
either, but—now surplus, outside. What
isn't, although it must have been once, me
but now is rented or is lent to me,
is paid as wages for the "work" I therefore "love." I do,
I "love" the work that lets me like the world,
"love" the indenture that I call my "gift,"
almost as much as simply fear
the blinding loss of it.

"Loving" in truth, I take its shape.
(But what that is I won't be first to know.)
Sir Richard Burton, on his first exposure to
boy prostitutes—to, in fact, the "execrabilis
familia pathicorum"—wondered why
whole boys commanded almost twice the rental

of young castrati.—"The reason proved to be
that the scrotum of the unmutilated boy
could be used as a kind of bridle for directing
the movements of the animal."

                                        To have,
as excess, the thing you might as easily not have
propels you forward with the impulse from behind,
the place you cannot see but others can.

Tonight—it's past 11—in the graveyard of St. John's,
Chinese the mysterious and Humby the mysterious
play at a game like this one.
The tough old whipper-in, catamite and familiar
to tough, ancient Lord Twytten
lends to the drugged, imperious boy his seamed
voice, his leathery obstetric hand,
at . . .
          Oh, at what?
                                        At "sex."
                                        At a game of *horses?*
Can that be true?
                                        Another Henry I know
says that the thrill, for him, in watching
video porn in the bar's back room's
the irrepressed consciousness that somebody
somewhere (Upstate?) is busy wracking brains
to figure what on earth it is that men
can Do Together.
                                        It is strange:
the way the art of our necessity
     makes precious, the vile things—
the finger's-breadth by finger's-breadth
     dearly bought knowledge
of the body's lived humiliations,
     dependencies, vicarities
that's stitched into the book
     of The Sexualities, wasteful
and value-making specificity.
(*The Warm Decembers*, chap. 8)

The third and last of the misrecognitions that I locate in this tableau of spanking-and-poetry is the one in which, as it seems to me, many of the most important elements of a narrative poetic, as well as of a sexuality and of a complicated gender identity, are to be found. The sexual politics of the ass are not identical to the sexual politics of the asshole. But they are not totally separate either, and neither (vide Freud among others) are the physical excitations to and of the contiguous areas.[16] Where a lively anal eroticism has come into being, whether resulting from or merely associated through metonymy with the tableau of spanking, it is likely to be marked in girls *or* in boys by many of the same problematics that have already yoked together the experiences of spanking and of bowel training: shame, rhythmic control, and prohibition, notably. At the same time, the same anal eroticism will propel the subject into an area of our culture where the gender dimorphism of discourse is almost un-thinkably extreme.

Put simply, the "problem" of men's anal eroticism (through the "problem" of male "homosexuality" to which it is representationally central although, needless to say, far from identical) is not only *an* important and meaningful theme in Judaeo-Christian culture but arguably inextricable from modern Western *processes of meaning* (social and economic as well as gender meaning) through and through.[17] This almost never means good

16. See, for instance, Freud's "Character and Anal Eroticism," in *Collected Papers,* vol. 2 pp. 45–50. In general, I am assuming the relation of buttocks to anus, as erotic and as representational zones, to be a varied and negotiable one, dependent on a reading of various features of the environment of meaning in which they or their representations are activated.

17. Of course, this is not the place in which that argument will actually be made; for this, see my *Between Men.* To explain the sudden bristling of quotation marks in this sentence: neither anal eroticism nor male homosexuality is intrinsically (outside of a homophobic discourse) a *problem,* naturally, in my view; the quotation marks around "homosexuality" signify the modern (rather homophobic) use of the word to represent the entire problematic of male homosocial desire/homophobia/homosexuality. As for the mutilating condensation, to some extent reproduced in this essay, of the entire range of male homosexual desire and identity to the single, "representationally central" thematics of anal eroticism, I should make explicit its double status here. My reflections here record finding that condensation simply where I have found it: first, in the self-contradictory homophobic obsessions of the culture at large, and signally for me in the nineteenth-century English culture where I largely read; and second, in the individual desires, needs, obsessions, and identifications early knitted together by one prehensile consciousness from piecemeal elements of that self-contradictory homophobic culture. That the resulting fabric of identity and sexuality should be spangled with distorting and dangerously incoherent nodes is no surprise; the intended work of this essay is neither to render invisible, to validate, nor to tighten those nodes, but quite the opposite, to explicate them patiently enough that whatever representative value they may have can be put at the service of a project of disentanglement.

news for men's anal eroticism, but it always means news—or more accurately, news always means this. The situation with women is almost exactly the opposite. Although there is no reason to suppose that women experience, in some imaginary quantitative sense, "less" anal eroticism than men do, it can as far as I can determine almost be said as a flat fact that, since classical times, *there has been no important and sustained Western discourse in which women's anal eroticism means*. Means anything.

Aside from the well-rehearsed though controversial asymmetries between the cultural salience of (the regulation of) male homosexuality and that of lesbianism, there is the further obvious asymmetry, more important here, that female anality hasn't, to put it mildly, the representative relation to lesbianism that male anality has to male homosexuality.[18] In fact, one of the few topoi in which the female anus ever becomes sexually visible is that of a woman's "being used as a man," as an anally receptive man or a man who is being raped. (The sexual act is, in this topos, invariably seen as degrading to the woman, when not presented simply as punishment.)[19] The more assertively "heterosexual" of the available topoi is the (pseudo)metonymy by which women's *genital* receptivity is described as "ass," as in "a piece of." What can be said about this usage is, in the first place, that it does indeed display the linguistic traces of the male homosocial structure whereby men's "heterosexual desire" for women serves as a more or less perfunctory detour on the way to a closer, but homophobically proscribed, bonding with another man: the phrase itself, for instance, is never addressed to a woman, but only "used behind her back" to another man. But in the second place, what is true of the punitive "using her as a man" topos is also true of the invariably expropriative and contemptuous

---

18. Though interestingly, it is almost only in contemporary lesbian writing (of a particular, "politically incorrect" libertarian stamp) that there is anything informative or engaging at all about women's anuses and their pleasures. See, for instance Pat Califia, *Sapphistry: The Book of Lesbian Sexuality* (Tallahassee, Fla.: Naiad Press, 1983), pp. 52–53, 110–13. It is with lesbians, too, that much of the original and liberatory thinking on fat has originated. See, for instance, Judith Stein and RaeRae Sears's broadside, "Fat, Lesbian, and Proud" (Coralville, Iowa: Fat Liberator Publications, n.d.); or more recently, the lesbian-affirmative and altogether wonderful *Radiance: A Magazine for Large Women*.

19. See, for example, the punitive climaxes of V. S. Naipaul, *Guerrillas* (New York: Knopf, 1975); and Richard Fariña, *Been Down So Long It Looks Like Up to Me* (New York: Dell, 1983). This train of thought runs very deep, even in women: presenting "Everything Always Distracts" (see infra) to a group of other feminist poets, for instance, I found the assumption that being buggered *had to mean* being degraded, and male-homosexually degraded in particular, to be so ineradicable and unexaminable that it very quickly became impossible for us to communicate with each other about the poem.

"piece of ass" topos: namely, that neither one even pretends to name or describe (never *mind*, value) the anus as a site of *women*'s active desire.

One must all too readily admit that there is, in the multiple discourses of our culture, no aspect at all of women's sexual desire that has been described as such very richly or valued as such very highly. But many other sites of women's pleasure do *mean*, even when they don't mean well. The endless institutional adjudication of pleasures between the clitoris and the vagina, for instance, while most often oppressive in its effects, must still be said to have had a diacritical importance for the European discourse around women in the last century and a half. The phobic English Renaissance discourse of vaginal insatiability, and similarly the horrifying survival of clitoridectomy into the present, are, if nothing else, testimonials to meaning. If the enforcing discourse of fat has produced a situation in which a woman who confides "I was *really bad* yesterday" is 99 percent certain to mean that she ate something she enjoyed; if the potential of the breast for receiving as well as giving pleasure has been, with oppressive consistency, portrayed under the aspect of maternity; if the similar potential of cutaneous contact in general has been, with oppressive consistency, portrayed as exploitable infantilism (demanding hugs), as exploitable maternity (giving hugs), or as exploitable narcissism (wearing nice textiles)—still these sites and pleasures *are* everywhere in Western culture portrayed and given meaning. Even what is still the most universally repressed and mutilated of pleasure-taking organs, the foot, must be said to have a particularly high and consistent representational value, in proportion as its disablement and torture in women is repeatedly made a centerpiece of the semiotics of gendered visibility (fashion). None of this is true of anal pleasure or desire, in women.[20]

20. It is too early to tell whether blue jeans ads will signal a gap in this discursive blackout. The aesthetics and consumerism of the ass and the asshole aren't, any more than their sexual politics, identical; and if the pouted bottoms of the women in the ads *are* stressing anything besides their own surfacial curves, it is perhaps, frictively, those ungiving front seams. (Imaginably, even, it's the fullness and positive *size* of the vulva that's for a change being dramatized here, even if through a form of bondage.) Still, increasingly women's blue jeans fashion has to be read intertextually with gay men's: the complication *there* being that the jeans iconography that seems to be aimed at gay men is more often plain frontal, textures of touch-worn denim tensed over baskets in soft, emphatic triangular folds. Yet, again: it's oddly not always easy to tell the women (topless, muscular, shot from the back) *from* men, in ads that are seen interchangeably in *Vogue* and *GQ*; are we not, women and gay men, as audiences being invited to cross-identify rather fully, through a kind of tacit, sometimes ripplingly hilarious complicity of representational office? The mappings that may ultimately result for each of our active but colonized bodies are unclear—and crossed again, of course, by volatile forces and institutions, such as the medical, that can't be confined to the language of style.

What we have located, then, is a really quite large vacant space in our culture that presents a kind of lovely laboratory for the testing of a Foucauldian hypothesis, if for whatever reason we were interested in looking at it that way. How far can or will an already gendered and physically very localized desire swerve, how radically will it misrecognize itself, in its need to join a preexisting current of discourse through which to become manifest, to be fulfilled, manipulated, or even frankly repressed—to become, in short, meaningful? The answer is: quite far indeed. Almost far enough. But not without cost; nor perhaps without leaving a trace of its own particular itinerary; nor without the potential for changing, for better and worse, however minimally along with its own direction that of the discourse it joins.

A week, when last seen, short of her twelfth birthday, the poet Eve Kosofsky is sitting, Beatrix-like and bulky, in the bathtub when there's a knock on the door and in comes mother.

She sits down for a chat. It's Easter vacation in the public schools, so both of them are home for the day, but mother has a piece of school news (from where?) to share, to talk over. It's about the delectable Monsieur O. who, it turns out, won't be back after vacation. Well, he's in hot water. Been arrested. It's a long—

But these are, after all, after everything, two women who really know and pretty much trust each other. It's not just a long story. Does Eve know that there's such a thing as men who love other men "the same way most men love women"?

No, I didn't know that.

What's more—in a family whose verbal exhibitionism about sex was (is) matched only by the "voluntary" conservatism of their (our) practice—it was stunning to learn that there was something that serious that I could have *not* known. That it *was* so serious, apart from its consequences for Monsieur O., was something to be learned only over a space of time, time spent both sociably and alone. If my mother's exposition of the case can't have been exactly matter of fact, neither, it goes without saying, was it hysterical sounding. The word, the concept, of *entrapment* didn't enter her sketch of disciplinary enforcement in the men's room of a downtown Y, but the message that men and their institutions could get all in a lather about things that women among ourselves knew how to take more urbanely must, judging by its indelibleness in my own map of the world, have come through from the start, that day, with an exciting clarity. In fact, "urbanity" itself must have been, for reasons you can already gather (secular Jewishness, the Cold War era siting, the premium on the rendering unrecognizable of violence), a potent though not uncontested value in this family—and a badge of it, of course, the learning of French.

That a possible, highly charged differential of "urbanity" lived somewhere in the near neighborhood of the question of male homosexuality was, in any case, the narrative of those weeks. To have *retrospectively* to learn that what I had been absorbing as the gorgeousness of Monsieur O. had been, for months, visible to my mother as something that meant with an entire difference, as "too pretty," was (in a now familiar pattern) first humiliating then empowering to me, as I learned my way around both a homophobic language and an attitude of complicit and manipulative skepticism toward it. There followed a rapid series of discoveries about the faultlines of discourse—that none of my friends had been told this story or knew anything about this form of love, for instance; that the adults in their vicinity became anxious when I, with increasing avidity, told what I knew—and evidently not because the adults didn't know about it but because they all *did;* that the affair O. was literally the only thing that one was explicitly forbidden to talk about in school; that, perhaps most oddly, an entire well-known nonsense vocabulary of insult and superstition (not being "queer," not wearing yellow and green on Thursdays) had only been waiting for of all things *this* content to become meaningful, personal, sinister. This particular hermeneutic voyage is, in this homophobically cloven society, a distinctive and revelatory one. Almost everyone makes it in one form or another, but people who for whatever reason—most typically, of course, our own sexual self-discovery—are impelled to make it with urgency and rapidity bear its distinctive marks very similarly for life. Nothing—no form of contact with people of any gender or sexuality—makes me feel so, simply, *homosexual* as the evocation of library afternoons of dead-ended searches, "wild" guesses that, as I got more experienced, turned out to be almost always right. Why, when I ask the *Britannica* about the crime of Oscar Wilde, does it tell me about "offenses under the Criminal Law Amendment Act," nowhere summarized? If information is being withheld (and to recognize even that is a skill that itself requires, and gets, development), must it not be *this* information? I don't know whether there can be said to be for our culture a distinctive practice of "homosexual reading," but if so, it must surely bear the fossil marks of the whole array of evasive techniques by which the *Britannica,* the *Reader's Guide,* the wooden subject, author, and title catalogs, frustrate and educate the young idea.

Along with, at any rate, *my* practice of homosexual reading—a well-taught skepticism about the representative adequacy of language, consorting perhaps not oddly with a pressing sense that there was something somewhere else for it to be adequate to, and with a (to me now) most imposing deferral of the question what any of this had to do with *me*— there was developing something else too, which I did not at the time think of as a practice of homosexual writing. I didn't, I suppose, only because I didn't need to, it was so obvious. For very simply a sudden urgency to write narrative poetry, after my twelfth birthday, was coextensive with, was the same as, one or another plot of male homosexual revelation. Lawrence of Arabia, David and Jonathan, *The Man from U.N.C.L.E.*, Roger Casement, the Round Table, an avant-Girardian reading of *Jules et Jim:* the materials one had on hand were motley enough, but it seemed clear that if one wanted (and suddenly I did) plot, *this* was plot. "Lawrence Reads *La Morte Darthur* in the Desert," *aet.* 14, starts, in the familiar Untermeter:

> Having killed
> Six Turks
> Personally, in style,
> I will enjoy
> Extreme quietness
> For awhile.
>
> They shot
> (Briefly) at our last
> Flag. It sighed
> And exhaled around
> The wound, and died
> As the wind died,
>
> And fell. Six
> Hundred camels
> Tramped its bellying.
> This incident,
> Neither vital nor
> Particularly sullying,
>
> Ended a certain
> Revolution
> Of colours. From
> That battle, I
> Without the shame-rag
> Of chivalry, come

As frank as Arthur
After the fall
And free as Adam.
So are we all
A nameless equation
Reduced to Latin.

I have a body . . .
(1964)

The years after age twelve, in which Oedipus-wise I was feeling, with
both a charmed directness and an authentically monumental self-ignorance,
my way into what I still take to be a mother lode of highest-grade narrative
and epistemological meaning in this culture, were also years of depression.
(A will-to-live, per se, has seldom in me been more than notional, often aggressively absent: its place taken, when it is taken at all, partly by an also
aggressive will-to-narrate and will-to-uncover, each with a gay male siting. That this is a vulnerably off-centered psychic structure—dangerous to
itself, but, potentially, homophobically dangerous as well—is clear to me.
The *clarity* of that particular danger is, I sometimes think, the most homosexual thing about me.) The depression of these teen years, at any rate, I
survived through passionate and loving relationships with—have I mentioned this?—women.

I have spent—wasted—a long time gazing in renewed stupefaction at
the stupidity and psychic expense of my failure, during that time, to make
the obvious swerve that would have connected my homosexual desire and
identification with my need and love, as a woman, of women. The gesture
would have been more a tautology even than a connection. Yet it went and
has still gone unmade. In among the many ways I do identify as a woman,
the identification as a gay person is a firmly male one, identification "as" a
gay man; and in among its tortuous and alienating paths are knit the relations, for me, of telling and of knowing. (Perhaps I should say that it is not
to me as a feminist that this intensively loaded male identification is most
an embarrassment; no woman becomes less a woman through any amount
of "male identification," to the extent that femaleness is always (though
always differently) to be looked for *in* the tortuousness, in the strangeness
of the figure made between the flatly gendered definition from an outside
view and the always more or less crooked stiles to be surveyed from an
inner. A male-identified woman, even if there could thoroughly be such a
thing, would still be a real kind of woman just as (though no doubt more

inalterably than) an assimilated Jew is a real kind of Jew: more protected in some ways, more vulnerable in others, than those whose paths of identification have been different, but as fully of the essence of the thing.)

No, about this what is abashing, actually, though telling, is how excruciatingly long it took me to make the connection that you have anticipated these twenty pages—that I gazed raptly away from for twenty years[21]—and more: that, to change metaphors sharply, what was keeping that child's seat glued to the epistemological bucking bronco of a more than transsexual identification was, numbly and blindly and almost simply, her attachment to her seat. In the handsome noun phrases of Charlotte Brontë, "what throbs fast and full, though hidden, what the blood rushes through,

---

21. But then why choose, of all moments, this (AIDS-marked) one to stop not knowing, when even the *New York Post* is eager to sensationalize "the unseen seat of life" as exactly "the sentient target of death"? Not that one always chooses, but I like the view of Joseph Sonnabend, a doctor doing AIDS research in New York. "One should take an aggressive view. The rectum is a sexual organ, and it deserves the respect that a penis gets and a vagina gets. . . . It's terribly important to actually do this, because anal intercourse has been the central activity for gay men and for some women for all of history. . . . One should celebrate the act, but indicate that there is an epidemic which has stopped . . . this activity, an important part of one's life, and that, hopefully, there are circumstances in which unprotected anal intercourse will again become possible between two partners, in time" (*New York Native*, 7–13 October 1985, pp. 22–23).

That was the, perhaps chiefly self-protective, footnote with which I meant to ask and then foreclose this question in the fall of 1985, when this essay was first drafted. Although I still (1986) admire the Sonnabend interview, there actually is, I now see, a much truer and simpler answer to the question I asked then: if, as far as I can remember, the misrecognitions around the anal eroticism sketched in this essay were entirely blind to themselves, as misrecognitions of a particular erotic localization, until just that moment of medical terrorization and cultural enforcement, mustn't that change of consciousness be connected with some intimacy to the fact that—for instance—the *New York Times* had for the preceding year made virtually daily use of the theretofore unthinkable phrase "anal intercourse"—and, at least latterly, in connection with heterosexual as well as male "homosexual" (the *Times* would not during this period use *gay*) sexual practices? One thinks of the brutality of the Wilde trials, which most visibly shut down many lives and cultural meanings, but which also, in vastly less foreseeable ways, opened up new ones and altered irreversibly the flow of those that remained; similarly the early identification of AIDS (as it has occurred in the United States) with gay men and their lives and sexuality is bound before it changes to have sharp and unpredictable discursive effects along with the predictable ones that involve, so far, the numbing din of loss and repression and scapegoating intermitted with nothing more refreshing than courage. An increased pungency and sexualization, across the population at large, of the sense of a female and male anus almost at the same instant zoned fragile and fatal, is perhaps unlikely to prove the most consequential or long-lived of the unpredictable effects (already, for instance, only three-quarters of a year after first drafting this essay, I feel at a considerable distance of dread and mourning from the buoyancy of its sayable sense of anal pleasure); but the unpredictability is genuine and with it the possibilities, for some, of at least certain yet unclassified imaginings.

what is the unseen seat of life and the sentient target of death"[22]—whatever it makes its abode in the physical and symbolic body, it transforms with a hot and elastic rage to mean.

I am, I must say, too "interested" in this fact to be willing to keep up the work of presenting to you in near-simultaneous translation a reassuringly (angrily?) pre-palimpsested image of the struggles, the vulnerabilities, the blindside of my own language about it. Too tired by this process or tangled in it to work any more, right now, on keeping my ass covered. I do see, however, as I know you do, that the move from the childish "lyric" tableau of shame, exhibition, and misrecognized fury to the more adult "narrative" pseudolinearity of differentiated gender and identity formation has represented not so much a subsumption as an acting out of those earlier dynamics. The expense, rhetorically, of spirit involved, the arduous labor of embodiment required, in "the finger's-breadth by finger's-breadth" construction of meaning around a site of meaninglessness, necessarily in turn leaves "undefended"—leaves to its own often strange devices—another battle flank, here probably the surviving childish armory of verbal sadomasochism. I should think that as the vastation of meaning surrounding women's anality propelled one leap of misrecognition, across gender, in a male homosexual identification, it probably at the same time fortified something else, a stubborn *siege* of misrecognition, in an unshakeableness of the sadomasochistic identification with the punished child. And pointed attention to the one as thematics will necessarily, perhaps, set the other defensively in motion as structure. The prospect gives one pause.

> "I feel convinced,"
> wrote Trollope, "in my mind that I have been
> flogged oftener than any human being
> alive. It was just possible
> to obtain five scourgings in one day
> at Winchester, and I have often boasted
> that I obtained them all."
> "Looking back over half a century,"
> he added (with the doggedness
> of the adult he made himself from the child
> whipped far beyond the reach
> of "perverse" transformation at any psychic
> exorbitance, of pleasure to

22. Charlotte Brontë to W. S. Williams, in Clement Shorter, *The Brontës: Life and Letters,* 2 vols. (New York, 1908), vol. 2, pp. 127–28.

himself, to his tormentors even, from
what Dr. Middleton calls the Grecian
portico of a boy),
"I am not sure whether the boast is true."

Left by his mother *in the mother's place*
his murderous father's bloated punching bag
("knocked me down with the great folio Bible")
—even yet, loathly afterbirth, to learn somehow
in his excessive thirties
a possible desire, possible and infectious,
for stories in which bouncy rationalists
are let however briefly or tragically to receive,
to embody with a sibylline humility
and briskness, the afflatus of the Bureau
itself—postal inspector facing down
the postmark forgery! entails unwound!
solicitors who know solicitors
who know the law on heirlooms! —and clerks,
clerks by the score, into whose equivocal sails
may breathe or may not breathe the zephyrs of ambition,
competence, real engagement;
against a background where another
figure is let horrifyingly fail
in the common work of giving face
and shape and color to some want.

It isn't always possible. I fear it for myself.
I see it in the body of Beatrix in the bath.
(*The Warm Decembers*, chap. 8)

It will be appropriate then if I end not with Beatrix and a narrative but
with Eve and, of all backward-looking things, a lyric, one from the period
of "Lost Letter," where the meaningless female grasp *has* meaning, but
only perhaps in being read through the blind bourgeois scriptural space of
domestic adultery, and only, if inadequately, in the voice of a man.

Everything Always Distracts

Oh Eve, help me erase those nastily scenic
afternoons with the goddamned objects
in the goddamned motel room, with both your and my

goddamned beauty; with me—your beloved—
grim, baffled, jaunty, looking
(as they say of gynecologists) in the pink,

which to us means the folded tissue of blood,
and you, dear naked girl, with the disposal of
this red explanatory lapful:

that's not our love, which is pure voice
and also a steady touch in an inky room,
making a grown man want to think

his eyesight is a costly adult disease.
Your voice, mooded and languid under my voice,
too soft, not quite continuous, not quite

your own in the penetrated dark
touching and instructing my uncertain one, which is
more simply the riddled voice of sexual desire

and, afterwards, of unsleeping tristesse
reminds me a little of the touch of writing
to the reading it inhabits, trying to sustain.

(I know you think I'm being fancy, or just flat.
Wait, though, I've got more for you.) If
it finally happens, if we discover

a night we can spend together, a night to make good
what so far is only the raging sift of the detail
of impatient arousal, it won't be more

our own than other nights. Everything always distracts,
not least on the aired and inky sheet
of our intentness. Take, say, this blind instant:

the one where your teeth clamp hard on feathers:
you're fighting tears, while down the broad half-gleaming
back the raised ass is wedged conclusively

open—and that's me. I'm fond,
identifying with your delight, of these
rooted oppositions: you being shut and open,

oral sadistic and anal masochistic,

face down and bottom up, and I
half in half out, stretching you, and taking

my own sweet, my own needy time about
going either way. I like standing in,
in unconscious magic, for your pay,

your turd, your baby; but more, I'm scared
by the scalding rush to the eyes, the rush to finish—and your
resistance to it. Scared by your pain and my

infliction of it. Scared, though, mostly by
your grasp on this now weary, infused rag
and your greed for spending me like this, but beyond

this moralized landscape of (I admit) both our desire,
your own desire extends unmet
and unprovocative. If the night finally

comes when you and I, one sleepless darkness
mimicking another darkness, penetrate
from room to room, or into breathless

room, I needn't wonder if your voice
hollows under mine, sounding delicate, or absent,
the glutted body of that voice being here.
(1975)

Amherst, 16 June 1986

"A Poem Is Being Written" was originally conceived for the 1985 "Poetics of Anger" conference at Columbia University. Part of the motivation behind my work on it has been a fantasy
that readers or hearers would be variously—in anger, identification, pleasure, envy, "permission," exclusion—stimulated to write accounts "like" this one (whatever that means) of their
own, and share those.
The essay is dedicated to my brother, David Kosofsky.

DIVINITY: A DOSSIER

A PERFORMANCE PIECE

A LITTLE-UNDERSTOOD EMOTION

Michael Moon and Eve Kosofsky Sedgwick

EKS: This is a dream I had a couple of years ago. I was shopping for clothes for myself at a store that was nominally Bloomingdale's. I was dubious about whether they would have any clothes that would be big enough for me, but a saleswoman said they did, adding that rather than being marked by size numbers, each size-group of clothes was gathered under a graphic symbol: over here, she said, were the clothes that would fit me. "Over here" referred to a cluster of luscious-looking clothes, hung on a rack between two curtained dressing rooms. The graphic symbol that surmounted them was a pink triangle.

I woke up extremely cheerful.

MM: My love of opera as a protogay child growing up in rural Oklahoma in the fifties had at least as much to do with the available "visuals" as it did with the music—opening nights at the Met photographed in living color in *Life* and *Look* and on television, featuring befurred and bejeweled divas, usually fat, radiating authority and pleasure, beaming out at cameras from the midst of tuxedoed groups of what I remember one of the slick newsmagazines of the time calling "hipless" men. I was struck by the strangeness of that locution even when I read it at age eleven or twelve; like so many other bits of knowingly inflected pseudo-information about adults, their bodies, and their mystifying sexualities, all I could figure out about what it meant for a reporter to call an elegant group of men in evening clothes "hipless" is that it must be another bit of code for doing what was called at the time "impugning their masculinity." It was a deep fear of mine as a twelve-year-old boy putting on pubescent weight that after having been a slender child I was at puberty freakishly and unaccountably developing feminine hips and breasts. My anxieties on this count made me a fierce discriminator of the prevailing representational codes of bodies and body parts, but everything about this urgent subject seemed hopelessly confused and confusing. Why was John Wayne's big flabby butt taken as yet another sign of his virility while my aging male piano teacher's very similarly shaped posterior was read as that of a "fat-ass pansy" by some of my nastier agemates? What was the difference between a hermaphrodite—a figure still presented in freak shows at the local county fair in my childhood—and a male movie star like Victor Mature who was considered hypermasculine despite his overdeveloped and to my and many other childish eyes quite feminine-looking breasts? Was a man supposed to have hips or not? What regimen of diet, exercise, and character-building could possibly produce the apparently unattainable ideal of right-sized and -shaped male hips on my seemingly out-of-control body—a body that was supposed to be neither "hipless," i.e., gay, nor "fat-assed," i.e., gay?

For many gay men, as for such diverse modern avatars of male sexual and social styles as Byron and Wilde and Henry James and Marlon Brando and Elvis, dramatic weight gain and loss have played a highly significant, much remarked but almost completely unanalyzed part in the formation of our identities. One happy aspect of the story of my own and many other gay men's formations of our adolescent and adult body images is that the fat, beaming figure of the diva has never been entirely absent from our *imaginaire* or our fantasies of ideal bodies; besides whatever version or versions of the male "power-body" of the seventies and eighties we may have cathected, fantasized about, developed or not developed, and, in our time, pursued down countless city streets, the diva's body has never lost its representational magnetism for many of us as an alternative body-identity fantasy, resolutely embodying as it does the otherwise almost entirely anachronistic ideal, formed in early nineteenth-century Europe, of the social dignity of corpulence, particularly that of the serenely fat bourgeois matron.

EKS: Catherine Gallagher has written[1] on the complex representational functions of the image of the large human body in political economy after

---

1. Catherine Gallagher, "The Body Versus the Social Body in the Works of Thomas Malthus and Henry Mayhew," *The Making of the Modern Body: Sexuality and Society in the Nineteenth Century,* ed. Catherine Gallagher and Thomas Laqueur (Berkeley: University of California Press, 1987), pp. 83–106.

Malthus. By Gallagher's account, Malthus in 1798 inaugurated a representational regime in which the healthy working body both continued, on the one hand, to function—as it had for millennia—as a symbol and prerequisite for the health of the social and economic body as a whole; but at the same time the same substantial and hence procreative individual body began on the other hand, through the newly activated specter of overpopulation, to represent the constitutive and incurable *vitiation from within* of that same economic totality, as well. After Malthus, Gallagher concludes, "a general sense of the body's offensiveness spreads out" from the large body "and permeates the whole realm of organic matter."[2]

The labor of concentrating and representing "a general sense of the body's offensiveness" is not a form of employment that will seem archaic or exotic to large women in modern American society. It permeates the mise-en-scène of my dream, the store where "I was dubious about whether they would have any clothes that would be big enough for me," whose implicit tension and dread must be resonant for almost any fat woman in this culture. The confrontation of the complex labor of representing offense with the female homosocial marketplace of gendered visibility—the materialization of a fat woman in a clothing store—lights up the works of a pinball machine of economic, gender, and racial meanings; at the same time as it is likely to register on the steeled body itself as insult, concussion, ejection. To that woman the air of the shadow-box theater of commerce thickens continually with a mostly unspoken sentence, with what becomes, under capitalism, the primal denial to anyone of a stake in the symbolic order: "There's nothing here for you to spend your money on." Like the black family looking to buy a house in the suburbs, the gay couple looking to rent an apartment, the handicapped high school kid visiting a barrier-ridden college in the Ivy League: Who and what you are means that there's nothing here for you; your money is not negotiable in this place. Distinct from the anxiety of never *enough* money, the anxiety that there won't be any roof for my head, food for my hunger, doctor for my illness—the more awful anxieties whose energy, however, at least knows how to be commandeered with a fluency just as awful into the capitalist circulation of meaning—this is instead the precipitation of one's very body as a kind of cul-de-sac blockage or clot in the circulation of economic value. My permeability to offensive meanings in such a situation comes, to follow Gallagher's argument, from the double and contradictory value exacted from my bodily representation. Visible on the one hand, in this scene, as a disruptive *embolism* in the flow

2. Ibid., p. 102.

of economic circulation, the fat female body functions on the other hand more durably (and through the same etymologic route) as its very *emblem*.[3] Like the large, dangerous bodies in Malthus, the modern fat female body represents both the efflorescence and the damaging incoherence of a social order, its function sharpened by representational recastings and by the gender specification, class complication, and racial bifurcation that accompanied the shifts from nineteenth-century European to twentieth-century U.S. models. Its consequence: that what I put on to go shopping in is the brittle armor of a membrane-thin defiance whose verso is stained with abjection.

MM: We have for some time been collaboratively compiling a dossier on a feeling or attitude we call "divinity." The presiding figure for these meditations has been, naturally, Divine, the late star of many John Waters films. As a huge man who repeatedly created the role of "the most beautiful woman in the world," Divine seems to offer a powerful condensation of some emotional and identity linkages—historically dense ones—between fat women and gay men. Specifically, a certain interface between abjection and defiance, what Divine referred to as "glamor fits" and which may more broadly be hypothesized to constitute a subjectivity of glamor itself, especially in the age of the celebrity, seems to be related to interlocking histories of stigma, self-constitution, and epistemological complication proper to fat women and gay men in this century. This combination of abjection and defiance often produces a divinity-effect in the subject, a compelling belief that one is a god or a vehicle of divinity.

The subjectivities from which we ourselves are enabled to speak are, it goes without saying, my own experiences of divinity as a fat woman, and Eve's as a gay man.

EKS: John Waters and Divine were a celebrated gay-man-and-diva couple who, until Divine's death in 1987, pursued powerfully mutually enabling careers in film and performance. That Divine, the eponymous diva in question, was not a woman but a biologically male transvestite is important to our project, but so is the way Waters's and Divine's respective body types play themselves out in the representational world of their films, writings, performances, and interviews. Like his mock-sleazy mustache, Waters's body is pencil-thin, what some would call "hipless." Divine's, by contrast, was that of a three-hundred-pound man not trapped in but scandalously

---

3. Embolism: cf. embolus: . . . < Gk *embolos* stopper = *em* + *bolos* a throw, akin to *bállein* to throw. Emblem: . . . < Gk something put on = *em* + *blema* something thrown or put, cf. *emballein* to throw in or on. Cf. also abject < L *abjectus* thrown down.

and luxuriously corporeally cohabiting with the voluptuous body of a fantasy Mae West or Jayne Mansfield.

MM: In the film and theater of the past two decades, as well as in the body of critical gender theory and performance theory that has arisen during the same period, transvestism has often been trivialized and domesticated into mere "crossdressing," as if its practice had principally to do with something that can be put on and off as easily as a costume. In fact, influential essays like Elaine Showalter's "Critical Cross-Dressing" have allowed transvestism to become *the* dominant image in feminist theory for the purely discretionary or arbitrary aspects of gender identity.[4] As such, it is sometimes treated as sinister—when men are seen as being empowered by a pretense of femininity they can doff at will, leaving their underlying gender identity and privilege untouched or indeed enhanced. Alternatively, a very similar understanding of transvestism can take on a utopian tinge: as a denaturalizing and defamiliarizing exposure of the constructed character of *all* gender; as a translation of what are often compulsory gender behaviors to a caricatural, exciting, *chosen* plane of arbitrariness and free play.

But the social field in which this universalizing, discretionary "theory" of transvestism gets mobilized is already structured by a very different, overlapping set of transvestite knowledges thereby repressed but by no means deactivated.

EKS: That some people can cross-dress convincingly and others can't.

MM: That some people's bodies make more sense to themselves and others when they're cross-dressed than when they aren't.

EKS: That some people get turned on when they cross-dress and others just feel at home.

MM: That cross-dressing crosses between public and private differently for different people.

EKS: That for some people, cross-dressing signifies their hetero-, and for other people their homo-, sexual identity.

MM: That the embeddedness of cross-dressing in routines, in work, in spectacle, in ritual, in celebration, in self-formation, in bodily habitus, in any sexuality, can vary infinitely from one person to another.

4. Elaine Showalter, "Critical Cross-Dressing: Male Feminists and the Woman of the Year," *Men in Feminism*, ed. Alice Jardine and Paul Smith (New York: Methuen, 1987), pp. 116–32.

EKS: That some people's cross-dressing is consistently treated as a form of aggression and responded to with violence.

MM: Divine's performances forcibly remind us of what so many treatments of transvestism require that we forget: that "drag" (as Esther Newton has suggested) is inscribed not just in dress and its associated gender codes but in the body itself: in habitual and largely unconscious physical and psychological attitudes, poses, and styles of bodily relation and response—not just on the body's clothed and most socially negotiable and discretionary surfaces. In addition, drag depends on, even as it may perceptually reorganize, the already culturalized physical "givens" of the body, among them ones—size, color, gestural scale—that may have near-ineffaceable associations of power or stigma or both. In stark contrast with the performance style of a relatively "respectable" "female impersonator" like Charles Pierce doing a characteristic turn as Carol Channing or Barbra Streisand in an upscale nightclub, Divine's fiercely aggressive performances do not conceal or disavow what a dangerous act drag can be, onstage and off. Nor do they gloss over how obnoxious many viewers find the act, especially if it is not hedged on all sides with half-truths about why performers "do drag" and why audiences enjoy it—e.g., it's merely a performing skill like any other; it's a classic theatrical tradition; it allows performer and spectator to let off steam without really challenging predominant gender and sex roles for either. Divine's "loud and vulgar" (to use her terms for it) drag style flings the open secrets of drag performance in the faces of her audience: that unsanitized drag disgusts and infuriates many people; and that it is not wearing a wig or skirt or heels that is the primary sign of male drag performance, but rather a way of inhabiting the body with defiant effeminacy; or, the effeminate body itself. And, finally, that it is just this conjunction of effeminacy and defiance in male behavior that can make a man the object of furious punitive energies, of gay-bashing threatened or carried out rather than applause.

EKS: Strangely, in fact, one of the most striking aspects of the current popular and academic mania for language about cross-dressing is its virtual erasure of the connection between transvestism and—dare I utter it—homosexuality. We might take as emblematic an article in the premiere issue of the late Malcolm Forbes' new magazine, *Egg*—which vindicates its claims to chic by featuring an interview with three downtown drag performance artists, an interview in which the word "gay" *is never spoken*. Or again, the business section of a recent *New York Times* ran an article (the front-page teaser headline was "Corporate Cross-Dressers?") about male CEOs of airlines, insurance companies, TV networks, and other established capitalist ventures who dramatize the work of business meetings by appearing in costume—including, as the *New York Times* puts it, "an intriguingly large number of top male executives who turn up in women's clothing."[5] As the subhead explains, corporate drag "can make a point, lighten a mood, or soften bad news." What it apparently cannot do is induce the acknowledgment of so much as the existence of divergent human sexual choice.

We have something of the same sense about most of the current theoretical and critical discussions of transvestism as we have about those uses of a nominally desexualized drag to oil the wheels of corporate business-as-usual. Uncharitably, one might say that gender theory at this moment is talking incessantly about cross-dressing *in order* never to have to talk about homosexuality. (This is the modus operandi of, for instance, the highly popular play *M. Butterfly*.) The usual alibi for segregating discussions of cross-dressing from issues of sexuality is the much-reproduced assertion that "the majority" of cross-dressers "are heterosexual men." We have a lot of trouble with this as an assertion. Survey research is notorious for turning up unexpectedly large concentrations of heterosexual men. Again, why do we suppose empirical research to be capable of telling us what a heterosexual man is, when nobody else can? And frankly, when was the last time any of us was invited to an earring party by a heterosexual man?

We have even more trouble with it as an alibi, however; by the very compartmentalization of cross-dressing between the hetero- and the homo-, it seems both to reduce the almost infinite array of cultural, personal, and contextual meanings cross-dressing can have, and at the same time to repudiate or traduce the profound historical linkages in Western European, English, Euro-American, and African American culture between drag performance and homoerotic identity formation and display. It is like pretending that the ancient music of Druidic rituals provided the roots of rock 'n' roll.

I have used the policial metaphor of the "alibi" to make a polemical point. There is a less accusatory way to put the problem, however, one that is surely nearer the spirit of the critics, some themselves gay and/or gay-loving people, who are busy theorizing transvestism in this odd conceptual vacuum. I think they think they *are* talking about homosexuality. After all, "everyone already knows" that cross-dressing usually at least alludes to homosexuality; "everyone already knows" that the surplus charge of recognition, laughter, glamor, heightened sexiness around this topic comes from its unspecified proximity to an exciting and furiously stigmatized

5. *New York Times*, Sunday, March 25, 1990, p. 33 (i.e., sec. 3, pt. 2, front page; "Corporate Cross-Dressers?" was the teaser headline on sec. 3, pt. 1, front page.)

social field. Critics may well feel that the rubric "cross-dressing" gives them, too, a way of tapping into this shared knowingness without having to name its subject; without incurring many of the punitive risks of openly gay enunciation in a homophobic culture; but also—advantageously as far as they can tell—without incurring the *theoretical* risks of essentializing homosexual identity, of presuming a given set of relations between gender identity and sexual object choice, or of ignoring how little coextensive the population of cross-dressers actually may be with the population of gay men or lesbians. But in that case, can't the tactic be innocuous or even useful? *Must* "everyone already knows" be misguided as the structuring strategy of a critical movement or moment? Only if "everyone already knows" is itself *already* the structuring strategy of a homophobic culture: specifically, of the culture's need to revivify itself constantly with the energies of gay experience, while maintaining a semi-plausible deniability about the gay history and sexual specificity of that experience. But that does, of course, describe the status quo exactly. Our culture as a whole might be said to vibrate to the tense cord of "knowingness." Its epistemological economy depends, not on a reserve force of labor, but on a reserve force of information always maintained in readiness to be presumed upon— through jokey allusion, through the semiotic paraphernalia of "sophistication"—and yet poised also in equal readiness to be disappeared at any moment, leaving a suppositionally virginal surface, unsullied by any admitted knowledge, whose purity may be pornographically understood to be violated and violated and violated yet again each time anew, by always the same information in fact possessed and exploited from the start. The "knowingness" most at the heart of this system is the reserve force of information about gay lives, histories, oppressions, cultures, and sexual acts— a copia of lore that our public culture sucks sumptuously at but steadfastly refuses any responsibility to acknowledge.

MM: The most nightmarish versions of this infinitely iterable violation-by-revelation tend to cluster around legal scenes: scenes like the trials of Oscar Wilde, but just as much like the hearings in the U.S. Congress that have led to literally genocidal prohibitions against spending federal funds on AIDS education materials that exhibit any tolerance for the existence of homosexual men. The scenario of these public denudations scarcely varies: they consist simply of the articulation, in so many words, of sex acts, which is to say names of parts of bodies, supposedly unique to or characteristic of gay men. William Dannemeyer's or Jesse Helms's catalogs depend heavily on the mantra word, the ever-new and ever-potent syllables "rectum," for instance (as if this weren't a fairly common thing to have and even to enjoy having). It depends also on the mantra word "homosexual" itself: I am thinking for instance of Dannemeyer's shockingly effective exposé to Congress last summer on the weird facts of gay sexual practice which, he says, "militant homosexuals do not want you to know"—including astounding assertions that the "average homosexual" has "homosexual sex" "two to three times per week," as well as that "other activities *peculiar to homosexuality*" include oral sex (sometimes mutual), anal sex, and the use of sex toys.[6] Two to three times per week! That such a snoozable statistic, when applied to, not sex, but "*homosexual* sex" performed by "*homosexuals*," would have power to ignite scandal and motivate legislation, is a testimonial to the inexhaustible reserve power of incredulity in our culture. And it is no mere paradox to suppose, as we do, that that reservoir of incredulity represents the invariable concomitant to the reserve of knowingness attached here to the simple adjective "homosexual." A powerful argument, we would say, for placing the project of gay/lesbian theory explicitly at the center of transvestite theory, and hence of gender theory more broadly.

EKS: Recent work by scholars like Marjorie Garber[7] demonstrates very valuably that the relished, taboo omnipresence in our culture of cross-dressing and trans-gender coding may well constitute the very possibility of gender coding at all. What this work does not consider—or at least does not take responsibility for enunciating—is that the rabid frenzies of public deniability are an inextricable part of the same epistemological system as the sophisticated pleasures of public knowingness—pleasures which such work itself richly indulges. The history to which this nominally historicizing analysis appeals (like the history that tacitly undergirds recent, more frankly dehistoricizing, psychoanalytic work on male subjectivity)[8] is never the history of sexuality, of changing and overlapping homo/heterosexual definition, of homophobic oppression and homo-affirmative resistance.

Now the gay specificity omitted from such accounts is undeniably a problematic, perhaps necessarily incoherent concept—but it has been all the more explosively potent across every space of our culture, for that. Indeed, the economy of "knowingness," far from deconstructing that in-

6. Rick Harding, "Sex Education in Washington, D.C.: Dannemeyer Talks Dirty on the Floor of Congress," *Advocate* 534 (26 September, 1989): 10; quoting from the *Congressional Record*, 29 June, 1989; emphasis added.
7. We refer to material in Marjorie Garber's book, *Vested Interests: Cross-Dressing and Cultural Anxiety* (New York: Routledge, 1992), on "cross-dressing and cultural anxiety."
8. The "Male Subjectivity" issue of *Differences* (Fall 1989) epitomizes such approaches.

coherent concept and its essentialist underpinnings, instead reifies it by silent presumption. Histories, like the one constructed by Garber, of gender play and gender transgression need also to bear explicit relation to explicitly problematized histories of gay specificity, identity, oppression, and struggle. Otherwise, in drawing a surplus value of pleasure and rhetorical force from cultural energies to which it never seems to consider itself directly accountable—in consolidating a community of "knowing" listeners who draw authority and cognitive leverage from allusion to gay communities and resources not always given the stabilizing dignity even of a name— such work risks being, not critical of, but isomorphic with the inflictive and demeaning enunciatory relations of the homophobic culture at large.

MM: When Glenn Milstead was in high school, his body and his effeminate way of inhabiting it infuriated people on sight: he sometimes needed a police escort merely to get to and from school. That he provided, at these moments of identity constitution and enforcement, an apt embodiment of the purely discretionary seems unlikely. When Waters renamed this high school friend of his "Divine," he both recognized Divine's affinity with the abject and apotheosized drag heroine of Genet's *Our Lady of the Flowers* and at the same time set the seal of a name on Divine's dangerous and exciting, though far from arbitrary, course of cultivating and valuing his brazen effeminacy as a primary component of his identity. This involved much more than becoming a "female impersonator"; the name on Divine's passport, we are told, was "Divine." Waters writes that as late as the time they were making *Pink Flamingos* Divine was unable to leave home without attracting violent attention. "His heavy lipstick seemed never to come completely off, so his lips and face were permanently stained a faint pink. Having little interest in his everyday male attire, he wore baggy one-piece white worksuits off the set, giving him the appearance of a demented, rather feminine garbage man," Waters says. "He looked even more bizarre out of drag than he did in."[9] If Glenn Milstead hadn't become Divine, what would he have become? Doesn't it devalue a creativity as deep as the bones and musculature, imperfectly delible as lipstick, and as painful as 300 pounds in high heels, to define it in the inconsequential terms of the free market in genders and identities?

When personal friends talk about Divine, they have a tendency to use the masculine pronoun in discussing the years before he took the name

9. John Waters, *Shock Value: A Tasteful Book About Bad Taste* (New York: Delta Books, 1981), p. 11.

"Divine," the feminine pronoun for periods through about the mid-70s, and the masculine pronoun again for the last decade or so of his life. This is a very rough division; and different people handle their pronoun usage about Divine very differently, not always on a chronological basis.

In any event, Divine in 1981 recollects having been "strange" but not necessarily "effeminate" as a child,[10] but being subject to much-increased gender harassment, by teachers as well as age-mates, in junior high and high school: the bizarre indignity, for example, of being placed in a girls' rather than a boys' gym class. After around age fifteen, new friends, the concept of "drag," and a new aesthetic that interfaced spectacle with filth intervened on Divine's life and identity, apparently feminizing her, declassing her from her comfortable origins, and propelling her toward a subjectivity of glamor. According to Jean Hill, Divine in the early 60s would claim in gay bars actually to have had a sex-change operation; though Divine later suggested she had only distantly considered having one, and only in the early 70s.[11] The last few years of Divine's life, marked by celebrity identity and success, seem also to have been marked by virilization: reflected in Waters's 1981 remark that "his drag fever has almost vanished," while he "lives quietly. . . with his longtime roommate, Phillip."[12]

This history is notable for how little support it gives to a conception of gender as *either* essential and unchanging, *or* free-floating and discretionary. In a person whose native attributes included a potent but very stigmatizable presence, the lack of a consistent lifelong core gender identity seems to have represented both a liability—part of the stigma itself— and a space for certain long-range negotiations and investments of creativity in what Erving Goffman refers to as "the management of spoiled identity." To the degree that Divine could negotiate gender, she used it as a way of hurling her great body across chasms dividing classes, styles, and the ontological levels of privacy, culthood, fictional character, celebrity, and, of course, godhead. Despite Divine's trajectory toward fame, it is not clear that *all* this mobility is best described as upward. It did not, perhaps, either, make her body steadily more intelligible to himself. Moreover, advantageous positionings for one of these ontological levels may well have been debilitating for others. At a certain active level of human creativity, it may be true that the management of spoiled identity simply is where experimental identities, which is to say any consequential ones, come from.

10. Ibid., p. 146.
11. "John Waters's Issue," *Pandemonium* 3: 23; Waters, *Shock Value*, p. 154.
12. Waters, *Shock Value*, p. 146.

To hypostatize these circumstances as either *compelled* or *voluntary*, in terms of either *work* or *play*, is, we would suggest, to give in to the available, tendentiously mystified metaphoric alternatives of *the machine* or *the marketplace*, and to do little justice to the exploratory reach of this particular body and the art that at least overlapped it.

EKS: We'd like to say a word at this point about the kind of intervention we are trying to make in the current uses of cross-dressing as a condensed emblem for the whole project of gender and sexual constructivism. Nothing could be further from our intent than to push backward against the constructivist trajectory in the name, or even in the direction, of an essentialism whose killing effects we take to have been amply documented. But we do fear that the choice of cross-dressing *as* emblem for the constructivist project may, along with the real progress it is still enabling (in, for example, the recent work of Judith Butler), also further a dangerous conflation of issues in the current framings of the debate on "constructivism" vs. "essentialism." Briefly, as regards gender and sexual identities, we fear a conflation of the question of what might be called phylogeny with that of individual ontogeny. The origin of this conflation probably has something to do with the double disciplinary genealogy of constructivism itself: on the one hand, through a Foucauldian historicism designed to take the centuries vertiginously in stride; on the other, through an interactional communications theory whose outermost temporal horizon is, in practice, the individual life span. The *phylogenic* question, which asks about the centuries-long processes—linguistic, institutional, intergenerational— by which such identities are or are not invented, manipulated, and altered, gets asked under the rubric of "constructivism" as if it were identitical to the *ontogenic* question: the question "how did *such-and-such a person come to be*," shall we say, gay rather than straight.

We see three problems with this tacit devolution from constructivism-as-phylogeny to constructivism-as-ontogeny, a devolution that seems to be facilitated by some current uses of the topos of cross-dressing. The first is simply the cognitive loss, a certain vulgarization involved in the ideational collapse. The second, as we have mentioned, is the frightening ease with which anything that our capitalist/consumer culture does not figure as absolute *compulsion* (e.g., addiction), it instead recasts as absolute *choice* through the irresistible metaphor of the marketplace. One, but only one terrible effect of this marketplace imagery is the right-wing demand that gays who wish to share in human rights and dignities must (and *can*) make the free-market choice of becoming *ex*-gays—an abuse of the constructivist analysis to which there absolutely must be some response stronger than the currently popular gay politicos' retreat into the abjectly essentialist, "We deserve rights and dignity because *we were born this way and can't help it*."

Finally, there is reason to be nostalgic for the exhilaration of that founding moment of gay liberation ideology, the moment when the question of gay ontogeny—"What makes Johnny queer?"—got dislinked, seemingly once and for all, from the assertion of the gay subject's claims on the resources and support of the society in which she must exist. The project of gay/lesbian liberation was possible *only* when the fascination, the consequentiality of the riddle of individual ontogeny had been shattered. So there is a clear (not to say prohibitive) risk in the reviving demand for *any* form of narrative in the ontogenic framework.

MM: In our attention to Divine we are especially interested in the part played in the process of her self-creation by celebrity itself—as a level of culture that refuses to keep its place as merely one level among many; as an ontological status that *dis*articulates the intersections among the person, the artist, the fictional character, and the commodity. Clearly, celebrity was part of what enabled the thereby-constituted Divine to make a certain, new sense of an impossible body.

EKS: What can a celebrity body be if not opaque? And yet what if the whole point of celebrity is the spectacle of people forced to tell transparent lies in public? We have already mentioned what we take to be a central chord in our culture of "knowingness"—the reserve force of information, the reservoir of presumptive, deniable, and unarticulated knowledge in a public that images itself also as a reservoir of ever-violable innocence. The economics of knowingness helps us ask new questions about the transparent lies that constitute celebrity, as well. Why do they have to be lies? And why do they have to be transparent?

> **MM:** I don't diet to look better, I diet to feel better.
>
> **EKS:** Now I'm sober and back in control of my own life.
>
> **MM:** Being a mother has made everything meaningful for me.
>
> **EKS:** I happen to be secure in my masculinity.
>
> **MM:** Reality is the greatest high.
>
> **EKS:** I'm taking my time looking for the right part.
>
> **MM:** I'm taking my time looking for the right woman.

EKS: I think that's why I [act]—to give people hope.[13]

MM: [Now I'm being Bette Midler]: I never explored the baths and I never went anywhere except the dressing room and the stage.[14]

EKS: I took a long look at who I really am, and you know what? I like myself!

MM: I had to spend two months watching people at gay bars to prep for this role.

EKS: I'm feeling better than I've ever felt in my life.

MM: I'm not a transvestite, I'm a character actor.

EKS: In a 1987 interview Divine says, "At this point, I can't help it if others have a lot of misconceptions about what I do, if they're not willing to believe I am a character actor and one of my characters just happens to be a loud, vulgar woman" (44). What needs—whose needs—are served by the construction of divine drag as one more job of work in the free market of the sartorial? And what needs are served when, in the chorus of voices saying and repeating this, no one believes it to be true?

MM: In the same interview, Hal Rubenstein of *Interview* magazine asks the question—obligatory in every celebrity interview—of typecasting. "If you want to be known as a character actor and want to get more male roles, doesn't perpetuating a drag character hinder you from changing the perceptions of ready-to-pigeonhole Hollywood casting people?" And he follows it up: "But is the typecasting all their fault?" And the next question: "Why is it that you are still haunted by your past, long after Goldie Hawn has shed her bimbo image and Raquel has doffed her *1,000,000 B.C.* loincloth?"[15]

Divine's rather inspired strategy for dealing with these questions is to persist in pretending to believe that they are all references, not to drag or to sexuality, but to a single, notorious scene in *Pink Flamingos*—the famous final shot, in which Divine, playing a triumphant Babs Johnson, aka Divine, "the filthiest person in the world," manifests her divinity by eating a mouthful of freshly laid dog shit. "It was designed to shock and make everyone aware of who we were . . . except, talk about having a hard act to follow!"

13. Interview with John Malkovich, *Interview* 19 (March 1990): 124.
14. Bette Midler, quoted in Hal Rubenstein, "Simply Divine," *Interview* 18 (February 1988): 51.
15. Rubenstein, "Simply Divine," p. 44; further citations in text.

"Why is it that you are still haunted by your past . . . ?" Divine's final answer: "Because they still want to know if I ate 'it.' It's so old. With everything that's going on in the world, how can that still be on anyone's mind?" (44).

EKS: Divine wants to talk about eating shit instead of about doing drag. A diversionary tactic that self-evidently can't succeed, it dramatizes the over-arching premise of the celebrity interview—how it stages the spectacle of divinity eating shit. How it ushers audiences onto an exciting and nause-ating scene of creation: the creation of the closet. Public scenes of self-misrecognition are a staple of human relations with the gods; a divinity with self-knowledge, on Olympus, on Sinai, couldn't be expected to have much in the way of world-creating or narrative-inducing powers. But the opacity of gods to themselves used to be a property of their own strength, rage, willfulness, lust, and jealousy. Now, a fascinated, vengeful calculus about who has the power to enforce or exact this spectacle energizes the public in the age of celebrity.

MM: Closet of sexuality, closet of size. But what can it mean—the closet of size? The pink triangle hovering over those big garments in Eve's Blooming-dale's dream? In one sense, in this dream, the relations of knowledge con-densed in the pink triangle (a penal marker attached in the first instance by force to a few men from the fear that their desire would otherwise remain subversively indistinguishable from the desire of the many around them) seem to be diametrically different from the relations of knowledge around those large female garments, garments that can only gesture at minimizing a stigma that could never be hidden because it simply *is* the stigma of visi-bility. But if it is really nonsensical to talk about the fat woman's closet, or if that closet is really destined to remain empty, then why in the dream was this riveting superposition of stigmatic images framed or flanked by the symmetrical pairing of identical, closed cubicles? What kind of secret can the body of a fat woman keep?

Gay people coming out to the people around us report, much more often than encountering a response of simple surprise, experiencing instead the relief that one's associates no longer feel entitled to act from the insolent conviction of knowing something about one that one doesn't oneself know. The closet, that is, seems to function as a closet to the degree that it's a glass closet, the secret to the degree that it's an open one. Nonsensically, fat people now live under the same divisive dispensation; incredibly, in this society everyone who sees a fat woman feels they know something about her that she doesn't herself know. If what they think they know is something as simple as that she eats a lot, it is medicine that lends this notionally self-evident (though, as recent research demonstrates, usually erroneous) reflection the excitement of inside information; it is medicine that, as with homosexuality, transforming difference into etiology, confers on this rudimentary *behavioral* hypothesis the prestige of a privileged narrative understanding of her *will* (she's addicted), her *history* (she's frustrated), her *perception* (she can't see herself as she really looks), her *prognosis* (she's killing herself). The desire to share this privileged information with the one person thought to lack it is more than many otherwise civilized people can withstand.

EKS: It follows from all this, however, that there *is* such a process as *coming out as a fat woman.* Like the other, more materially dangerous kind of coming out, it involves the risk—here, a certainty—of uttering bathetically as a brave declaration that truth which can scarcely in this instance ever have been less than self-evident. Also like the other kind of coming out, however, denomination of oneself as a fat woman is a way in the first place of making clear to the people around one that their cultural meanings will be, and will be heard as, assaultive and diminishing to the degree that they are not fat-affirmative. In the second place and far more importantly, it is a way of staking one's claim to insist on, and participate actively in, a renegotiation of *the representational contract* between one's body and one's world.

MM: In her fascinating essay in speculative history, *Another Mother Tongue: Gay Worlds, Gay Words,* Judy Grahn seems to suggest that gay women and men have, cross-culturally and transhistorically, shared through such roles as that of shaman the liminal and potentially transformative function of representing cultures to themselves.[16] It is hard to pin down, and no doubt quite variable, how discretionary the individual assignment to, or choice of, such ceremonial roles in a given culture may be; what varies even more clearly across economies and cultures are the ideological, material, moral, relational networks of support and reabsorption for these highly volatile and apparently necessary enactments. For instance, in contemporary culture, certainly, gay men since about the 1950s have added to their stereotypical late nineteenth-century work of representational preservation and renewal (in art, literature, opera, design, couture, etc.) the labor of representing the straight male body to itself: the straight

16. Judy Grahn, *Another Mother Tongue: Gay Words, Gay Worlds* (Boston: Beacon Press, 1984).

man who looks for the culture's most influential images of him will be looking at Rock Hudson, James Dean, Montgomery Clift, Cary Grant, Marlon Brando, the ephebes or the bodybuilders in *GQ*. Such work is, in our culture, often materially rewarded but at high cost: our puritanical rage against representation itself, manifest in the age of television in a contemptuous orgy of trivialization and in the multiple rubbishing of the lives of our representers, exacts as well from each of these totemic men the mutilating tribute of a public self-misrecognition. The names and images that haunt the straight man in search of his image, at the same time, will be those in which a not altogether dissimilar content of personnel and imagery is explicitly labeled *gay*.

EKS: It is a simpler story but perhaps not an utterly different one when my sister who is deliberately starving herself, under the real or imagined gaze of some man or some other woman, looks in the mirror in the morning and the body that she thinks she sees confronting her is—mine. *No one* would choose the labor of embodying to this woman, to the society that has created this astigmatism in her view, the shame and anxiety of her and their (and my) own economic exploitiveness, physical greed, sexual subjugation, mortality, and unloveliness. No one would choose it, but it is labor—wearing, wasting, perhaps necessary, in any case *exacted* labor, and must be seen and valued as such.

What some women would choose, and do, in particular those of us with the resource of our various lesbian communities: One response—a possible one, but it should not be a compulsory one—to this exaction, is to attempt, acknowledging and sharing the heritages of older women, immigrant women, African American women, to struggle actively with the given bodily code for *material* accumulation until it surrenders, as well, some of its immemorial meanings of the accumulation of spiritual, physical, sexual, and intellectual power. In some of us there is the project, even the necessity, to try to embody this further transformation *in* ourselves and *for* our sisters—and brothers—who, willy-nilly, see their bodies in ours. Again, however, if that is a form of creativity it is also a form of labor, and in this culture a dangerous and fragile one. The support for it from all of those whose open and covert representational needs it serves, itself requires new forms and new embodiments.

MM: *As* a form of representational labor, the fat woman's work of emblematizing the circulatory embolisms of a culture might be said to fall into the economic category, not of either production or reproduction, but rather of waste management. The way in which human fat, and especially fat-gendered-female, has represented economic accumulation and waste in post-Enlightenment Western culture is a complicated narrative. Briefly, by the mid-nineteenth century, bourgeois Euro-American women were rigidly subject to a "sphere" ideology that appeared to make their economic position absolutely distinctive, and distinctively that of material consumption, their mercantile husbands' circulatory role manifesting in them as sheer absorption.[17] Thus, when caricatural figures for what Catherine Gallagher refers to as "the fatted body of circulation" would come to be looked for in the bourgeoisie, it was to a very specifically gendered fat body that these meanings were most ineffaceably attached.[18] Dickens is close to the modern nerve with his authentic loathing for the fat female body: the utter and inalterable inability to be forgiven, of precariously middle-class fat women

17. By the time of Mayhew and thereafter, in England, an avoirdupois dimorphism of class had come to be the accompaniment of the class-marked differential of gender relations. The reader of Dickens or Trollope would not routinely mistrust a professional or mercantile-class adult man on the sole basis of a certain *embonpoint* or, as it's tellingly called, corporation— while the prodigal baggage allowance of a Mr. Pickwick would look, as Catherine Gallagher's examples show, perfectly depraved on a Sam Weller, or that of a Cheeryble brother on a Newman Noggs or Mortimer Knag, whose proper morphological heritage is the "puny body of production." But whereas for people of the working classes the sexual divisions of labor did not become so marked as to effectually sequester women away from the urban labor marketplace—so that, to go back to *Nicholas Nickleby*, we can be offered the exemplary figures of Miss LaCreevy, Madame Mantalini, and the Infant Phenomenon, seen at their paid labors of miniature painting, dressmaking, or acting, and all as scrawny, as visibly stunted by overwork and undernutrition as the bantamweight men around them—for the bourgeoisie, the more radically gendered body fantasy increasingly obtained.

18. Lucy Snowe dramatizes this shift in interpretive perspective when, in chap. 19 of *Villette*, she rambunctiously pretends to misperceive a seventeenth-century Rubensesque odalisque as a portrait of conspicuous consumption in a contemporary realist idiom: "It represented a woman, considerably larger, I thought, than life. I calculated that this lady, put into a scale of magnitude suitable for the reception of a commodity of bulk, would infallibly turn from fourteen to sixteen stone. She was, indeed, extremely well fed: very much butcher's meat—to say nothing of bread, vegetables, and liquids—must she have consumed to attain that breadth and height, that wealth of muscle, that affluence of flesh. She lay half-reclined on a couch: why, it would be difficult to say; broad day light blazed round her; she appeared in hearty health, strong enough to do the work of two plain cooks; she could not plead a weak spine; she ought to have been standing, or at least sitting bolt upright. She had no business to lounge away the noon on a sofa. She ought likewise to have worn decent garments: a gown covering her properly, which was not the case: out of abundance of material—seven-and-twenty yards, I should say, of drapery—she managed to make inefficient raiment. Then, for the wretched untidiness surrounding her, there could be no excuse. Pots and pans—perhaps I ought to say vases and goblets—were rolled here and there in the foreground. . . ." Charlotte Brontë, *Villette* (Edinburgh: John Gray, 1905), pp. 333–34.

like Flora Finching, seems to suggest a literal-minded *imaginaire* of political economy where the gibbous flesh of such women might be carved directly from the narrow shanks of the smaller bodies—bodies of children, of the poor—in which Dickens saw himself.[19]

EKS: Anne Hollander, in *Seeing Through Clothes*, gives an illuminating account of the aesthetic and material involvements of the shift, after World War I, from a fat to a thin norm for the well-off female body.[20] In an increasingly abstractive economy, the Dickensian revulsion at female size as a phobogenically literal image of exploitive accumulation results, not of course in any revolution in the exploitive economic structures themselves, but in an extravagantly sublimatory semiotic reassignment: not her bodily opulence but her bodily meagerness comes to be the guarantee of the woman of substance. The Duchess of Windsor's gyneocidal pronouncement, "You can never be too rich or too thin," marks the absolute boundary of this semiotically ambitious, not to say psychotic, sublimation of use value in exchange value.[21] By contrast, under this economy of radical sublimation, the fleshy female body is catastrophically declassed; so that a large woman of any class or race will now feel more at home among the round faces brown, white, and black of Brooklyn, than being stared at over the circumflex cheekbones of Madison Avenue.[22]

Of course, however, there is no such thing in any culture as a simple reversal of meaning. The shift of thinness from being a lower-class to an upper-class female signifier, and vice versa of fatness, had among its mediators one especially powerful discourse—the medical—whose structure of knowledge, at once highly elastic and relentlessly *naturalizing,* insured that what emerged from the shift of bodily meaning was not a clean and newly

---

19. When the plumpness of a Mrs. Pocket or a Mrs. Jellyby comes actually to verge on the infanticidal, Dickens's fat phobia swims into sudden binocular focus with the hideousness of his marital obsessions: the fat-emblazoned scandal that, in their *sexual* function, they will not cease from productive labor.

20. Anne Hollander, *Seeing Through Clothes* (New York: Viking Press, 1978).

21. An illustration of the more pluralist aesthetic attaching to use value would be the blues formulation that contrasted women "built for comfort" with the also attractive alternative of those "built for speed."

22. The few large-size clothes that are available for purchase owe to this their penitential meaning for professional-class women: the most portable marker of contemporary class privilege, the wearing of care-intensive natural fibers, is rigorously excluded in the design of all but the most extortionately expensive big garments, so that extent of female body gets manifested in tracts of class stigmatization and cutaneous discomfort, seldom in surfaces expressive of private delight or public display.

inscribed slate of role assignments, but instead a palimpsest of fragmentary meanings, inscribed in a biologistic narrative that can only take itself for the most direct commonsense, but whose actual gaps, overlays, and semi-erasures spell out a much less enabling rebus: a pattern of discreditation and impossibility for the female body of any class and race and of any size. In some of its meanings, the medicalized discourse of fat simply reproduces, in a disguised and hence nonconfrontable form, the direct reading of political economy taken over from Malthus: eating high on the food chain, a diet rich in animal fats or in commodity crops like sugar, spices, and coffee, is ecologically greedy *and in addition* unhealthy; the moral fervor of *Diet for a Small Planet* gets deposited unquestioningly into the account of *The Bloomingdale's EAT™ Healthy Diet*—a transfer enforced by the raging background din, in women's lives, of that further and crazier imperative whose sick-making instruction is, *diet for a small swimsuit.*[23]

MM: Interestingly, it is in the nascent unfolding of a movement much younger than gay/lesbian liberation, namely the fat liberation movement, that the liberatory moment of ontologic *dis*linkage is currently being enacted. New science (much of it being done by gay men scientists) is finally getting around to demonstrating the commonplace—discursively valueless so long as it was spoken only by women, by fat people—that fat people do not actually eat more than thin people. Whether or not *because* of this "scientific development," at any rate, the issue of *being fat* is able to be, even today is being, severed thrillingly (though still with an unstanchable incompleteness) from the moralizing discourses of greed or the medicalizing discourses of "eating disorders"—to be established instead in the assertive, anti-ontogenic space of an emergent identity politics. That the politicized insistence on a willed agnosticism about individual *causes,* the anti-ontogenic crux moment in fat liberation, rhymes so closely with the analogous moment in gay liberation, records a profound and un-

23. And the coercive incoherences of this palimpsestic discourse insure that when, for example, dieting itself begins to be, as it is now being, labelled as a pathological, addictive "disorder" of life-style, that damning diagnosis of thinness or of noneating does nothing to budge the damning diagnosis already delivered on fatness and on eating. In a culture where the compulsory may become visible only as a manifestation of the individual will, medicine allows the concept of addiction to play a pivotal role; it insures that any behavior, any condition of being, is subject to discreditation on the grounds that, while it appears to be an exercise of will, it is, in fact, *compulsive.* (I didn't just make up *The Bloomingdale's EAT™ Healthy Diet*—it's a real book, whose cover explains that the word EAT, *not* here a real word, is meant to be decomposed and read instead as an acronym—EAT dissolving into the (trademarked) Effective Appetite Training.)

acknowledged historical debt. It might point as well to the political need for a historicizing, phylogenic, anti-essentialist construction of size indeed already under way in such work as that of Hillel Schwartz as well as in our present work.

EKS: The ontogenic dislinkage of fat is, however, the furthest thing from the obsessive mind of John Waters; indeed, it is his abolute refusal of such a move that makes the center of gravity of his inimitably hefty thematics. In a late-capitalist world economy of consumption, the problematics of waste and residue, hitherto economically marginal, tend increasingly to assume an uncanny centrality. The concept of "ecology" itself, with its profoundly, permanently destabilizing anthropomorphization of the planet as a single living body, emerged in the 1970s much less from the question of how to feed its inhabitants than from that of how to contain or innocuously to recirculate their wastes. At the level of the disciplines surrounding the supposed individual body, the recent strange career of cholesterol in the medical and public imagination suggests that to the conflict between virological and immunological body models, dramatized in discourses around cancer and AIDS,[24] there must be added a muted but potent third term involving not just cardiovascular medicine but the discipline that has come to be called garbology. The issue (in many ways a startlingly new one) of the very viability of our planet has emerged as the need, not merely to limit waste, but—no doubt you'll pick up on the paradox involved—to eliminate it. Which, paradoxically again, can only mean to consume it.

One consequence of these developments has been that the Enlightenment Western fantasy imperative of the hygienic has, not come to an end, but come under increasing and transformative stress. At the moment when Mary Douglas can construct an *anthropology* of hygiene, at least the transcendent self-evidence of the expulsive, projective hygienic project must be nearing its close.[25] If an ecological system includes no "out there" to which the waste product can, in fantasy, be destined, then it makes sense that the meaning-infused, diachronically rich, perhaps inevitably nostalgic chemical, cultural, and material garbage—our own waste—in whose company we are destined to live and die is accruing new forms of interpretive magnetism and new forms, as well, of affective and erotic value.

MM: From his earliest feature-length film, unsurprisingly titled *Mondo Trasho* (1969), onward, John Waters has been the filmmaker who most in-

24. See Cindy Patton, *Inventing AIDS* (New York: Routledge, 1990).
25. Mary Douglas, *Purity and Danger* (New York: Praeger Books, 1967).

sistently offers erotic, problematizing images, and performs foregrounded acts, of otherwise taken-for-granted economic processes of consumption, absorption, and waste. Waters is said to claim that his April birthday means he was "born under the sign of Feces"; for that matter, just being named John Waters might conduce to a toilet bowl mentality. Waters's project does not involve any simple, merely paradoxical reassignment of equations between filth and value, although there are moments of his work that could be taken in isolation as doing so. Rather, through a series of metonymies around the body of Divine, he explores one materialized displacement after another: food as clothing, clothing as bodies, bodies as food, bodies as waste, waste as food—and only in these contexts, waste as value.

EKS: *Food as clothing, clothing as bodies:* for only one example, the scene in *Pink Flamingos* in which Divine orders a steak from the butcher, then, after (as the screenplay has it) "carefully survey[ing] the store for detectives, unwraps the steak, and sticks it up her dress and into her crotch. A look of bliss comes over her face when she feels the cold steak against her warm flesh."[26] The bloody steak, diverted from the cash-lubricated path that was to have circulated it as if automatically from living animal to prime cut to dinner to nutrition to waste, instead recovers its materiality as a substitute for the absent panties; as an allusion to the messy, uncomfortable, and fascinating sanitary napkin; as a stimulation to, and representation (as "meat") of, the (absent) female genitals; ditto to and of the impermissibly present male ones, which it also, presumably, veils, bloodies, and comforts; and as a literalization of the fat-phobic fantasy of food that, rather than sublime and desublimate through the channels of digestion and fat formation, simply applies itself directly to the thighs. By the time the steak is served as dinner, Divine explains that the reason this rag of meaning is so "delicious" is that "I warmed it up when I was downtown today in my own little oven" (TT, p. 40).

MM: *Bodies as food:* Then there is the delicate matter of occasional outbreaks of cannibalism among the personnel of Waters's Dreamland. Two policemen caught spying on Babs Johnson's birthday party in *Pink Flamingos* get torn limb from limb and snacked upon, while *Desperate Living* ends with the "victory feast" of a huge platter containing the wicked "Queen Carlotta, cooked and garnished" (TT, p. 177). And hot dogs and marsh-

26. John Waters, *Trash Trio: Three Screenplays:* Pink Flamingos, Desperate Living, and Flamingos Forever (New York: Vintage Books, 1988), p. 20. Further cited in text as TT.

mallows get roasted over Vera, an unlucky rival-in-filth of Divine's, when she gets torched in the unproduced screenplay *Flamingos Forever.*

EKS: *Bodies as waste:* An intimate, sometimes almost warm Grand Guignol sense of bodies as offal permeates Waters's films. Exemplary moment: a woman in *Desperate Living* (originally supposed to have been played by Divine) gets a penis transplant to surprise her woman lover, but instantly and uncomplainingly cuts it off on finding her grossed out by it instead. According to the screenplay, her lover then "screams, picks up sex-change penis off the dirty floor and throws it out front door. A mangy dog on the street immediately eats it" (TT, p. 165). A flasher in *Pink Flamingos* supplements his natural endowment with a tied-on turkey neck. Again, the ingenue in *Mondo Trasho* gets her feet cut off by a mad doctor, who replaces them with huge, misshapen feet pulled out of a plastic bag kept in his toilet bowl. For the rest of the film she seems wistfully ashamed of her ugly new feet, but at a dangerous moment they're almost as good as ruby slippers: she clicks her heels together and is transported, magically, from a hog farm in Baltimore to a shopping center in Baltimore.

MM: *Waste as food:* the opening credits of *Desperate Living* probably say it all on this score, "superimposed over an elegant dinner table. A pair of black hands sets the table and pours some wine. Another course is served—this time a boiled rat heavily garnished. A pair of white hands with knife and fork enters the frame, cuts rat, and spears hunks of rat meat. Finally, the fork is set down and a rat bone is placed on center of plate" (TT, 96). The dog shit, whose swallowing in *Pink Flamingos* certifies Divine's divinity, is only the most dramatic synecdoche for Waters's constant deroutinizations of the "ordinary" circulatory relations by which the large body enters, figures, and incorporates the economics of its surround.

EKS: The process by which circulation and signification interrupt each other is generic as much as it is thematic. It is in the genres that he salvages, recycles, or parasitizes to make showcases for the egregious figure of Divine that Waters participates most pointedly in an ongoing gay cultural project as well. The best emblem for this might be *Mondo Trasho*'s momentary *hommage* to Kenneth Anger (as distinct from its format from beginning to end, which parades an indebtedness to the techniques of *Scorpio Rising*). The movie begins with its ingenue, Mary Vivian Pierce, boarding a bus and pulling her reading matter out of her purse: Anger's *Hollywood Babylon,* recognizable as such by the lush cover picture (taken of course before her decapitation in a car crash) of the head and bust of Jayne Mansfield. In his autobiographical book *Shock Value*, Waters says of the film, "I wanted to make real trash this time, and I knew Divine would make the perfect star. We both idolized Jayne Mansfield, and since Divine was getting quite heavy, we agreed she could play the perfect takeoff of a blond bombshell" (p. 54). Divine has, at this moment in the film, yet to make her appearance. But the chiasmus by which the gorgeous, soon-to-be-severed head of the first Hollywood star to make, on her own account, a total profession of celebrity, inverting the supposed hierarchy of product and publicity with an unprecedented candor—the process by which that head stands in for one gay male director being honored by another who trademarks his own films, or figures his own spare off-camera body, with the framed, shamed, celebrated, finally (in this film) disemboweled, irrepressibly overripe body of Divine, uncovers a mise-en-abîme of relations between production and waste—of relations, too, among gender, sexuality, and the ontology of genres.

MM: Waters's films of the 70s (*Multiple Maniacs,* 1970; *Pink Flamingos,* 1972; *Female Trouble,* 1974; *Desperate Living,* 1977) derived from and contributed to a whole range of novel and newly visible social and cultural practices of the time. Gay men and lesbians were often in the vanguard of the development of these practices, many of which were considered grotesque, obscene, perverted, decadent, and/or déclassé by the mainstream. A partial catalog of these practices might include radical drag street theater (the Cockettes, Hot Peaches); glitter or glamor drag as a style of mass performance (Mick Jagger in the film *Performance,* David Bowie, Kiss, etc.); punk subcultures and their sympathy with and frequent enactment of violence, self-mutilation, "bad attitude," hostility to polite hypocrisy and bourgeois social forms in general; "exploitation" films, especially the subgenres of motorcycle, women criminals, women's prison, and low-budget horror; collecting kitsch and spectating related forms of "bad taste" in tacky locales: striptease joints, live sex-show acts, exhibition wrestling, gambling joints, cockfights, pit bull fights; and, perhaps most importantly, various cults of "sleaze"—"anonymous" sex in baths, porn bookstores and theaters, peep shows, backroom bars; enacting "perverse" sexualities, e.g., S&M, exhibitionism, golden showers, and scat; performing, filming, and consuming pornographic films, videos, and magazines.

It is hard to imagine John Waters's films' ever getting made without the occurrence of the Stonewall rebellion in 1969 and the subsequent dissemination of a wide variety of gay identities and "life-styles" through what soon became an elaborately reticulated network of media representations and leisure markets (urban ones especially but by no means exclusively), representations in the media, and gay political organizing; yet the relation of Waters and his films to recognizably and avowedly gay-affirmative, anti-homophobic political and cultural practices is far from simple or straight-forward. This is true not only of Waters' work but of that of most of the gay male artists, filmmakers, and performers from whom his work derives and with whom he deserves comparison: Jack Smith, Andy Warhol, Charles Ludlam. Like their work, the spaces in which Waters's films occur com-prise a whole series of communicating and interlocking closets—spaces of concealment and disclosure, of avowed, denied, or misrecognized identi-ties of gender, body type, race, class, and sexuality.

EKS: One of the high costs of supporting the systemic series of inversions that Waters performed on bourgeois culture and "sleaze" subcultures is that male-identified gay men, middle-class by definition in Waters's construc-tion of things, can figure only as abject villains in the plots of his films. I'll take *Pink Flamingos* as an example. In that film, Divine plays Babs Johnson, an outrageous woman who, along with her demented son Crackers, her sleazily glamorous "traveling companion" Cotton, and her mother, Edie (Edith Massey), claim to be "the Filthiest People Alive." Connie and Ray-mond Marble, a greedy, social-climbing, uxorious, and extremely scrawny straight couple, peddle drugs to schoolchildren and run a business kidnap-ping young women, getting their servant Channing to impregnate them, and selling the babies to lesbian couples. The reliance of the Marbles' raptly infatuated, mutually narcissistic coupledom on the compulsory maternity of enslaved women and on the need of lesbians marks them sufficiently as guardians of the Name of the Family, however they may plot—vainly—to replace Divine and her entourage as "the Filthiest People." Now Waters is a gay man,[27] and so in an important sense was Divine, but neither of them

27. In his writing and interviews that we have seen, Waters has not chosen to describe himself in precisely these words, but has also not demurred at others' doing so. For example, when Chris Bull asks him in a recent interview "What role has being gay played in your develop-ment as a filmmaker?" his response is gently to despecify: "I've said this before: I don't think being gay or straight makes one good or bad at all. It definitely makes you accept a little more and see all sides of any question and be more compassionate toward people. I think every-one has what society sees as some neurosis, and everybody has secrets; everybody has things that nobody knows. And I think that's very, very interesting." Chris Bull, "No More Dirty Waters," *Advocate* 549 (24 April 1990): 30–35, quoted from p. 34. But he does tell stories about, for example, taking a male date to a lunch at the White House (John Waters, *Crack-pot: The Obsessions of John Waters* (New York: Vintage Books, 1987), pp. 74–75). Still, "I

performs that role as such in Waters's films; as the writer and director of his films Waters is the ubiquitous but invisible, and consequently disembodied, source of much of what the films enunciate, and Divine plays not a gay man who does drag performance (as, for instance, Harvey Fierstein does in *Torch Song Trilogy*) but a drag-monster version of an autonomous and obnoxious underclass woman at the center of an experimental family—in short, a female gang leader.

MM: The only character in *Pink Flamingos* who is recognizably a gay man is Channing, servant to the villainous and "assholic" Marbles and hated jailer and inseminator of the young women they hold captive. At first there seems to be almost no difference between employers and servant; Channing seems to be a perfect servant, in every way a mere extension of the will of the Marbles.[28] Channing's distaste for heterosexual intercourse, even as part of his job as the Marbles' lackey, is the first sign the film gives that he is going to be pressed into service as not only the Marbles' but in a sense the film's lackey and its emblematic gay man. The thorough abjection of this only secretly, never openly, defiant character culminates in the scene late in the film when Connie and Raymond Marble return from torching Babs Johnson's trailer house and discover Channing dressed in Connie's clothes playing at impersonating her and Raymond to amuse himself. The scene is structured like an encounter between an abject boy discovered playing in his mother's clothes and his furiously hostile parents. The Marbles pour homophobic contempt on Channing; Connie slaps him repeatedly, and Channing weeps and ineffectually defends his behavior as "just playing" (TT, p. 49): He says, "Stop hitting me. I didn't do anything to you. I was just here by myself and I start feeling funny when I'm alone. Those girls are down there, don't forget. I can't stand being in the same house with them. I can hear them screaming and crying, and then I get all nervous—then I get these spells. I don't plan it, it just happens, and then, well, I think about my

---

don't want to ever categorize myself as anything, because then you can't sneak in; you can't mentally creepy-crawl" (*Advocate*, p. 35).

28. Significantly, it is in relation to his role as rapist-inseminator that his difference from them begins to emerge. Midway in the film, Channing chauffeurs the Marbles around Baltimore so that they can kidnap another female victim for their "baby ring." In the scene that follows, Channing drags the unconscious woman down into the Marbles' dungeon cellar and announces to the other woman already captive there that he's devised a less disagreeable way of impregnating the new victim. The actor who plays Channing then proceeds to simulate masturbation, and then to simulate artificial insemination with a hypodermic syringe (Waters singles this scene out in his 1988 preface to the published script of *Pink Flamingos* as one that now offends even him). [*TT*, "Introduction," p. ix.]

position, my social standing, just like you two do, and I just play. I make believe that I *am* you. I know it isn't reality, I know I'm really me" (TT, pp. 47–48). When Raymond then locks him in a closet and calls him a "closet queen," Channing begs from inside, "Please! Don't lock me in! I'll just stay here and be me while you're gone; I won't even think about being you."

When Babs/Divine and Crackers arrive soon thereafter to put a hex on the Marbles' house, they release Channing from the closet and he guides them downstairs to the pit, where Babs and Crackers free the captive women who take their revenge on Channing by castrating him. In the published version of the script, there is a still from the film of Channing lying on the floor of the pit, dead, eyes and mouth agape, trousers lowered, and "castrated" crotch caked with blood. This photograph is captioned "Channing's Just Deserts." Channing *is* a villain in the economy of the film, but he is also the only character whose recognition and embrace of his own abjection does not "redeem" him. I find the image in the film of the dead and castrated Channing a particularly disturbing one because it images precisely the widespread and violent homophobic fantasy—and, horribly, sometimes the reality—of a gay man's supposed "just deserts," a spectacular scene of total abjection: public castration and death.

In *Desperate Living* the same actor, who actually is named Channing— Channing Wilroy—plays another lackey, a member of the effetely vicious household guard of the wicked Queen Carlotta: versions of the hideously durable homophobic stereotype of the gay man as fascist and Nazi. Here, the fact that the gay guards are massacred as a group in mid-orgy seems to register the pressure of a genocidal wish against gay men, a wish that Eve argues in *Epistemology of the Closet* has been endemic in our culture for the past century, never more than it is today.

Do these films in their theatricalized castrations and executions of their only visibly gay male characters simply mirror this murderous homophobic fantasy? I want to argue that they do not. A second way of reading the image of Channing's castrated body in *Pink Flamingos* is as a figure for Waters's "cutting himself out" of the film as self-aware gay male author, and "cutting out" straightforwardly gay-affirmative representation altogether. Channing's murder and castration is supposedly his "just deserts" for his role of holding women captive *and*, as the Marbles would have it, for "borrowing" their identities, for "playing" them by impersonating them, for being both a drag queen, as Connie calls him, and a closet queen, as Raymond calls him. Such a fate has a strong resemblance to the anxieties Waters himself may have felt and wished apotropaically and symbolically to turn aside, by repeatedly and theatrically "sacrificing" the kinds of roles he casts

Channing Wilroy in. Waters, after all, has been a powerful exemplar of the practice he embodies in his films—that of "borrowing" or "trying on" and "playing" with the identities and bodies of abjected groups, particularly highly adversary types of women, gay men, and transvestites. Channing's lackey position suggests that Waters's "lack"—as director rather than star, as male-identified rather than transvestite gay man—is a centrally enabling one for the kind of project Waters's films represent.

EKS: Capture and castration, then, are important markers of Waters's authorial self-insertions / self-excisions in relation to the film-fantasy world over which he invisibly presides. His direct representations of gay men are one major focus for these effects, and his representations of the maternal are another. The Virgin Mary, especially in her role of Stabat Mater, the mother in extremis following her divine son to the scene of his ultimate abjection, is repeatedly a key figure in this process. The "mystery" of the Virgin, according to Christian tradition, is that she was both Mother and Maid. The role of the Virgin and the roles of Mother and Maid, translated into their modern bourgeois forms, mistress and maidservant, housewife and cleaning woman, comprise an unholy trinity of recurrent roles for women in Waters's productions.

Indeed, it seems as if the diminished prestige and visibility there of gay men as such may be systemically related, not only to the apotheosis of Divine, but to an almost explosive multiplication around her of strongly figured female bodies and personalities visibly grappling with the flux of spoiled identity, hovering between mother and maid. These include, in particular, Jean Hill, a 400-pound black woman, and the also huge Edith Massey, a gravel-voiced, gap-toothed, radiantly magnetic, declamatory, and probably retarded older white woman. Massey might seem at first glance to be anything but defiant, exemplifying only the abject side of "divinity," but such an impression fails to register how many of the prohibitions and exclusions constitutive of our culture's representational codes Massey defied by performing, and by allowing Waters to feature her as a "star." Massey's life before her Waters-engineered apotheosis was a typical one for many women of her underclass background: she grew up in an orphanage, passed in and out of prison, subsisted for a long time on the seamier fringes of show business, engaged in casual prostitution and frequently worked as a barmaid in Oklahoma, Chicago, Las Vegas, Florida, and, finally, Baltimore, in the waterfront tavern where she was "discovered"—a place where, according to Waters, writing in his noir mode, "drinks were fifty cents and *any* kind of behavior was tolerated." Waters also says that Massey's first response to his asking her to play both her-self—that is, a den mother barmaid in a waterfront den—and the role of the Virgin Mary in "The Stations of the Cross" sequence in *Multiple Maniacs* was to decline because, she claimed, she couldn't act. But Waters disagreed: "After seeing her in the bar, charming every type from drunken sailor to nodding junkie, I knew she was wrong" (*SV*, p. 180).

Waters perceives what he saw as Massey's characteristic behavior, her potential both as an actress and as a star, as simply "charming" socially in-tractable and transgressive types of people, but we might look more closely at the uses he himself found for her skill at manipulating, pacifying, and coping with supposedly incorrigible and potentially violent people. After being shot and almost killed in May 1968, Andy Warhol decided to stop cul-tivating druggies and the other kinds of socially common "psychos" who had frequented the Factory and to make himself unavailable to the very people from whom he claimed to have drawn much of his energy during his most spectacularly productive period in the mid-60s. Isolated by what he came to see as regrettable choice from the "crazy" and "sleazy" people with whom he had formerly surrounded himself, Warhol felt that he had cut himself off from the sources of the disturbing and disruptive percep-tions that had enabled his best work. When Warhol met Edith Massey in the early 70s he was impressed and delighted with her, taking Waters aside to implore him, "*Where* did you find her?" Waters, like Warhol, recognized the value of "trash" and "sleaze" as sources of much of the most powerful and engaging representational work of the time. As highly privileged pro-ducers of such work, they were probably more aware than most of us can be of how imperiled and potentially lethal (and how perilous to the privi-leged positions that enable appreciation and expropriation) are the social spaces—the streets, the drug and sex-business subcultures—from which that representational work ultimately derives. Massey's so-called charm, her ability to "cope with" the denizens of these worlds by drawing on hard-earned street smarts and not by invoking class privilege or "calling the cops," might well have looked enviable, if not indeed magical and somehow even divine, to the (in this sense) "lackey" consciousness of the differently situated, and consequently differently vulnerable, Waters and Warhol.

Edith Massey eventually achieved a Waters-influenced celebrity as a per-formance artist and doyenne of a Baltimore thrift shop called, inevitably, Edith's Shopping Bag. She died after *Polyester* was completed, but her loud, uninflected, achingly kind voice can still be heard on a Rhino Records an-thology of the worst recordings ever made, singing, almost unbearably, her theme song, "Big Girls Don't Cry."

MM: This economy of female proliferation results in, among other things, a thematized lesbianism often in the place where a gay-male possibility might have been broached. The climactic victory of the band of lesbian revolutionaries in *Desperate Living,* for example, must be one of the most resoundingly triumphant sexual-political moments of closure in popular representation in the decade after the Stonewall Rebellion; as we have suggested, the downside of this moment is that it is underwritten by what is perhaps the most spectacularly self-hating moment in Waters's films, the simultaneous representation of gay men as a fascist goon squad, "deservingly" cut down in mid-orgy.

The prolific spawning of Waters's divas results not only in a thematized lesbianism, but in the almost miraculous absence of male homosocial circulation of any recognizable kind; and most characteristically, in a chunky and funky Mariolatry. For example, Waters's first film narrative of Divine's divinity, the 1969 *Mondo Trasho,* is organized like a medieval saint's life straight out of the *Golden Legend,* crossed with the *National Enquirer.* The film represents Divine on a supposedly typical day, experiencing a series of shocks and disasters and reacting to them with the combination of abjection and defiance that in the course of the narrative render her "divine." The Virgin Mary appears to her three times; first in a laundromat, then in a snake pit insane asylum, and finally in the hog pen where she crawls to die at the end of the film.

EKS: The martyrology *Polyester* is the film that most literalizes the topos of Mother and Maid—and that makes the most of the potential that topos holds for deroutinizing the metaphorics of rubbish, as well. *Polyester* is the only film whose diegesis assumes that Divine's obesity makes her unlovable and powerless rather than magnetically irresistible. In this nightmarish— i.e., naturalistic—frame she is forced, herself, to embody the hygienic imperative, as an abased housewife, Francine Fishpaw, whose impossible dream is a normal and germ-free nuclear family. Through all the stations of her humiliation—her husband abandons her for his skinny secretary, her son is exposed as the psychopathic "foot-stomper" who has been terrorizing Baltimore, her daughter gets pregnant, her glamorous new boyfriend turns out to have been conspiring all along with his lover, Francine's *own* (need I say skinny) *mother*—through it all, her companion, comfort, and faithful friend is Cuddles (Edith Massey).

MM: "Oh, Franciney, *every*thing's gonna be *just fine.*"

EKS: An ex-cleaning woman who has been left a fortune by a grateful client after a lifetime of cleaning toilet bowls, Cuddles is the only one of

Massey's roles where she seems to "look retarded," paradoxically because the nouveau riche signs of prepette sportivity are also readable as stigma: the scraped-back-ponytail head as microcephalic, the sweater and hockey kilt as institutional.

Divine's numinous abjection encompasses, more than it diametrically opposes, the hygienic imperative—much as she and her son in *Pink Flamingos* had put a spell on the Marbles' house, inducing the Marbles' own ultrachic furniture to rise up and reject them by coating the chairs and couches with drool. *Polyester* theatergoers were issued "Odorama" cards containing various fragrances and stenches, to be scratch-and-sniffed at appropriate moments of the action (a pizza, a fart, etc.). All the scratching and sniffing puts any viewer, however, in the subject position of Francine Fishpaw herself. Though doomed herself to concentrate and radiate "a general sense of the body's offensiveness" as never before—through her bulk in the claustrophobic, Sirkian domestic space, through her hapless snoring, through the 5 o'clock shadow that keeps impending over her various chins—the endlessly meek and patient Francine is the most fanatical votary anywhere in Waters's films of Hygiene itself. Like an eight-armed divinity, every arm wielding an aerosol cleaner or deodorant, Francine lives by the projective fantasy of a hygiene that would re-naturalize (as The Family) a space from which production, circulation, excess, predation, and waste were alike evacuated to an outer, unimagined space thereby hypostatized as the vengeful Truth of nature. Accordingly, Francine/Divine, as Everyhousewife, is endowed with only one, spectacular talent: a prehensile and almost paranormal receptivity to offensive odor that causes her to spend much of the film darting heavily about her own house trying to catch up with her own flaring nostrils—snuffling noisily at bedclothes and the cracks of doors—wriggling uncontrollably as her nasal antennae tune in to invisible wafts—behaving, in short, like any scratch-and-sniffing animal in the world except *Homo sapiens domesticus nuclearus*.

Surely part of the reason Divine spends much of *Polyester* in olfactory overdrive is that this 1981 film of Waters's was his first to escape an X rating for the more easily bankable and more widely marketable R. Francine Fishpaw's flaring nostrils are a sign of, among other things, the now internalized censor hysterically sniffing out embodiments and enactments of filthy flesh, the primary business of the earlier films. Like them, *Polyester* has the generally paratactic form of a mock Stations of the Cross, a series of excruciating scenes of devoutly cultivated gross-out, but in it, as we have suggested, the gross-outs get domesticated, deodorized, and depilated as they get dematerialized. Of the rankly material circulation of "wasted" humans and—their chief sign—human waste in the earlier films, only a few desultory farts are allowed to linger.

It is by now impossible for many of us who are most interested in studying the joint career of Waters and Divine to make any retrospective assessment of their twenty-year collaboration without registering the pressure of AIDS on our interpretations. In his "Is the Rectum a Grave?" Leo Bersani interrogates Simon Watney's distressed recognition that under the dominant representational regime that has been constructed around AIDS, gay men's rectums have been figuratively posted with a DO NOT ENTER sign—"Premises Off Limits by Order of the Department of Health," as it were.[29] Nineteen-eighty-one, the year of *Polyester*'s release, was also the year that many people in and around urban communities in Western Europe and the United States began to register the scope of the threat and reality of AIDS. The rectum, previously the site and source of so much aggressively represented pleasure in Waters's work, *has* almost become a grave in *Polyester*. Characters in the preceding films had eagerly, even frenziedly, pitched their tents in the place of excrement, but no one seems to desire to do so in *Polyester*, least of all the newly hyperhygienic Francine. In *Polyester*, only the liminal appearance of the roseate buttocks of the pizza delivery boy in Francine's wet dream about "ordering out" momentarily suggests the possibility that despite AIDS the rectum may *not* have gotten resituated permanently out of sexual bounds as a site of erotic pleasure of many kinds, including scopic ones.

MM: The most conspicuous textual site for considering the centrality of the anus and the anal, experienced scopically or otherwise, in Waters's and Divine's work must be *Pink Flamingos*, which we hereby rename—or actually merely reaccent—*Pink Flaming O's*. As Naomi Schor does in her work on resituating the clitoris and the clitoral as central and informing rather than marginal detail, anyone interested in making anality a central concern of analysis must counter a pervasive epistemological bias in much psychoanalytic theory (as well of course as in the wider culture) in favor of the phallus and the phallic.[30] On the conventional road map of the body that our culture handily provides us, the anus gets represented as always behind and below, well out of sight under most circumstances, its unquestioned stigmatization a fundamental guarantor of one's individual privacy and

29. Leo Bersani, "Is the Rectum a Grave?" in Douglas Crimp, ed., *AIDS: Cultural Analysis, Cultural Activism* (Cambridge, Mass.: MIT Press, 1988), pp. 197–222.
30. Naomi Schor, *Reading in Detail: Aesthetics and the Feminine* (New York: Methuen, 1987).

one's privatively privatized individuality, as argued by theorists as otherwise different as British cultural materialist Francis Barker and French gay theorist Guy Hocquenghem. In closing, we want briefly to bring to bear on *Pink Flamingos* Bersani's question, "Is the rectum a grave?," as well as the question that D. A. Miller's recent work on Alfred Hitchcock's *Rope* and anality raises, "Is the anus a cut?," or, rather, "How, under what circumstances, and for whom does the anus, especially of the gay-male body, get represented as a 'cut,' as a sign of castration? In what other kinds of representational compacts may it figure?"[31] Both questions may call to the mind of the student of the history of sexual representation in the recent past a series of significant moments in film in the time since the making of *Rope* that take up the problem of figuring anal erotics between males in a number of modes, both tender and violent. In this talk, we can only point to a couple of instances. One would be Ken Jacobs's 1963 underground film *Blonde Cobra,* in which at one point Jack Smith presents his bared buttocks to the camera with a butcher knife handle placed to appear to be protruding from his "stabbed" anus while he cries in voice-over, "Sex is a pain in the ass. Sex IS a pain in the ass." The other would be the male-male rape scene which David Lynch is said to have filmed for *Blue Velvet* and then cut sometime before it was released—I have been told at the request of the actor Kyle MacLachlan, who played the character who was to be the victim of the rape, the boy Jeffrey. In the symbolic rape Frank (Dennis Hopper) carries out, he "kisses" lipstick onto Jeffrey's mouth, lip-synchs Roy Orbison's "In Dreams" to him, and then beats him unconscious. I have written elsewhere about how the uncanny effects of this odd ritual permit viewers of the scene to register at some level that a male-male rape is being represented without seeing the rape—or rather, a simulation of it—actually being enacted—a scene which might have been not only too much for Kyle MacLachlan's budding career but also too anxiogenic for many heterosexual-identified male viewers.[32]

EKS: There is nothing of the uncanny about the episode in *Pink Flamingos* that is most closely related to these scenes of real and mock anal rape between males. At Divine's birthday party a boy steps onto the performance platform, doffs his posing strap, lies down on his back, and throws his legs

31. D. A. Miller, "Anal *Rope,*" in Diana Fuss, ed., *Inside/Out: Lesbian Theories, Gay Theories* (New York: Routledge, 1991), pp. 119–41.
32. Michael Moon, "A Small Boy and Others: Sexual Disorientation in Henry James, Kenneth Anger, and David Lynch," in Hortense J. Spillers, ed., *Comparative American Identities: Race, Sex, and Nationality in the Modern Text* (New York: Routledge, 1991), pp. 141–56.

into the air. So far we have the very scene—that of a recumbent gay man with his legs in the air—that Bersani argues in "Is the Rectum a Grave?" is the one that in highly hystericized form unconsciously fuels such homophobic violence against people with AIDS as the notorious attack on three children in the town of Arcadia, Florida, in 1987. But instead of getting fucked as the viewer may expect him to, this young man in *Pink Flamingos* astonishes everyone, once his legs are in the air, by—we hardly have terms for what he does—beginning to "lip-synch" to a record by rapidly flexing his anal sphincter. The scene has a potentially powerfully desublimating effect on many other more conventional scenes of relatively highly sublimated negotiations around the anus of the male body, such as the lip-synching and beating scene in *Blue Velvet* into which anal rape is sublimated. The rectum is demonstrably not a grave nor the anus simply a cut in this representational scheme. This pink, flaming asshole not only makes an impressive show of something we think deserves to be called self-determination, it speaks, and indeed sings.

MM: And what does it say? Interestingly, in Waters's arrangement of things it gets its own song. The other goings-on at Divine's birthday party have taken place to the tune of a 45 that sounds like an inquisitive five year old; the only words to that tune are "Why 'n' why 'n' why?" sung over and over again. When the asshole starts "singing," the song changes to one that goes, "Mau-mau-mau, mau-mau mau-mau-m'mau." What is it saying? Should we be surprised—I suppose not—that it is talking baby talk? It is announcing, I believe, both that it is impersonating the mouth, and that it wants "mau-mau." Who or what, we may ask in response, is the asshole's mother?

Perhaps this anus is naming its mother at the same time it announces the object of its impersonation: the mouth, or "Mau-mau-mau." As it happens, no mouth, maternal or otherwise, appears to answer its call. As if in belated and displaced response to it, a few scenes later, when Divine and her son Crackers break into the Marbles' house to hex it, they cap the hex with an incestuous maternal blow job on the Marbles' couch. As Divine starts to go down on her son, they perform a verbal duet, operatic in its intensity, in which the divinizing effects of the defiant and abject behavior in which they are engaging emerges in explicit terms. Again, it should perhaps not surprise us that as Divine and her son begin to enact their parody transgression "curse," the dialogue waxes increasingly theological:

> CRACKERS: Mama! Mama! ["Mau-mau-mau"] I just thank God above I was lucky enough to be the soul that was placed in my body; the body of Divine's son! The body and blood of another generation of Divinity.

BABS: My only baby, Crackers! My own flesh and blood, my own heritage, my own genes. Let Mama receive you like Communion. Let Mama make a gift to you, a gift so special it will curse this house years after we're gone. Oh, Crackers, a gift of supreme mothergood [misprint for "motherhood"?], a gift of DIVINITY! (p. 61)

EKS: One of the most refreshing aspects of the representation of fat women in Waters's films (at least until *Polyester,* which in this respect does anticipate the cumulatively compromising effect of Waters's commercial success) is the resistance manifested by Divine and others, male and female, who play fat women in the films to being assimilated to the maternal role, as fat women commonly are in representational schemas of many sorts. When Divine is playing a mother in these early films, she is reliably a terrible one (she kills her grown daughter in *Female Trouble* for becoming a Hare Krishna) when she is not being a monstrous and therefore "divine" one, as she is in the scene just quoted. In these films, organ is represented as yearning for organ, more often than not along forbidden or at least thoroughly involuted paths, but desire that follows the line of familial roles gets ruthlessly inverted and shortcircuited; of familial feelings, only incest is respected.

MM: Divine as the Filthiest Woman in the World; Divine as sainted martyr of the dictum, Crime Equals Beauty; Divine as phallic Mother and Maid in a world that doesn't envision the joining of Father with Son. If some of these inversions are couched in familiar terms, I hope we have at least suggested how vast is the distance between them and any of the more perfunctory aestheticizations whose claim to a "subversive" political correctness is based on less searching experiments with materiality, identity, economic representation, and the flux of levels of culture.

If Waters's experiments are literal-minded, perhaps there is hardly an alternative way of refusing to take for granted how chunks of literality inject themselves into the circulatory system of symbolic consumption. One of the reasons we are eager to celebrate the literal-mindedness of Waters's early films is because of the dogged (you should pardon the expression) resistance that it seems to offer to a cultural economy of knowingness. If literal-mindedness can never be successful in disambiguating the status of the literal itself, at least it can be stubborn about injecting its emblems and embolisms into the circulatory system of allusive deniability.

We have had especially in our minds the contrast between Waters's sleazily refractory apotheoses, dubious and forever off-key as they are in their refusals and misrecognitions of gay male affirmation, on the one hand; and on the other, the suave and conventional religiosity, the unproblematized access to abstractive cultural authority, of hygienically post-Stonewall gay official culture. A foundational example of the latter might be Andrew Holleran's classic 1978 novel of post-Stonewall gay New York, *Dancer from the Dance,* with the slickly allusive comfort it draws from a whole lexicon of religious metaphor. The habitués of Holleran's discos "glisten[] with sweat like an idol around which people [kneel] in drugged confusion ... assuming the pose of supplication at some shrine";[33] they "sprawl[] like martyrs who have given up their soul to Christ" (p. 31); they "pass[] one another without a word in the elevator, like silent shades in hell, hell-bent on their next look from a handsome stranger" (p. 30); they pantomine "the ecstasy of saints receiving the stigmata" (pp. 30–31); they walk through a disco door and are "baptized into a deeper faith, as if brought to life by miraculous immersion" (p. 35); etcetera, etcetera. The unquestioning facility with which these images can be invoked as simply "metaphor"—can be invoked, for that matter, as simply "religious," as they surely could not be in the insistently graphic abjections, scapegoatings, body eatings and blood drinkings, "inspirations" (i.e., blow jobs), transubstantiations, and literal stigmatizations and divinizations of Waters's films—constitutes a virtual orgy of the trivializing but highly legitimating presumption that religion can be stably located, and innocuously exploited, at the sterile distance of "metaphor" from a subculture itself thereby stably located as a securely bounded topos.

We are struck also by Waters's incalculable distance from any authorized account of camp—depending as these have done from Susan Sontag to Andrew Ross on knowing presumptions about the difference between "depth" and "surface," or between levels of culture that in fact mutually constitute, block, circulate, and emblematize one another.

EKS: Each of us, Michael and I, with our different, overlapping, shared, and exchanging hungers, loves, loyalties, and pleasures, with our braidings and leapfroggings from nostalgic to futuristic cultural projects, with the mutual adventures of our spoiled identity, finds a rich supply of things in Waters's and Divine's work of a kind that often seems to be as scarce as it is precious: opulent images and daring performances that suggest the experiment of desires that might withstand the possibility of their fulfillment.

We especially relish—it was the scene, in particular, that finally brought together my own identification with and desire for this impossible, inspiring, ruined and experimental figure—Divine's mirror aria at the end of

---

33. Andrew Holleran, *Dancer from the Dance* (New York: Bantam Books, 1979), p. 27; further citations in the text.

Multiple Maniacs, after murdering her lover, Mr. David, and his new girl-friend:

MM (Divine [to a mirror]): O Divine, you're still beautiful. Nothing that has happened has changed that. . . . I love you so much! And you're still the most beautiful woman in the world! And now you're a maniac! O but what a state of mind that can be!

EKS: (Divine discovers the dead body of her daughter Cookie behind the couch; laments over it and then, extending herself on the couch, begins the following monologue:)

> **MM:** You're finally there, you're finally there, Divine.
> You don't ever want to go back now!
> Oh, Oh, Divine, you have to go out in the world in your own
> way now.
> You know it's all right!
> You know no one can hurt you.
> You know no one can even get near you.
> You have x-ray eyes now, and you can breathe fire!
> You can stamp out shopping centers with one stub of your foot!
> You can wipe out entire cities with a single blast of your fiery
> breath!
> You're a *monster* now, and only a monster can feel the fulfillment
> I'm capable of feeling now!
>
> *Oh* Divine, it's wonderful to feel this far gone!
> This far into one's own mania!
> I'm a *maniac!* a maniac that cannot be cured!
> O Divine, I am Di-*vine!*

EKS: Whereupon a fifteen-foot-long lobster shuffles into the room and attacks Divine.

We wrote "Divinity" fully collaboratively, through discussions over a period of about a year. (The listing of the authors' names is alphabetical.) The name attached as the speaker of any given section is seldom in a more than accidental relation to who originally wrote the section. "Divinity" was written for the conference "Discourses of the Emotions," at the Center for Twentieth-Century Studies at the University of Wisconsin, Milwaukee, in April 1990. A few pages of "Divinity" were originally part of a paper, "Labors of Embodiment," that EKS wrote for the MLA national convention in 1986.
Jonathan Goldberg and Hal Sedgwick were first among our many unindicted co-collaborators in "Divinity."

WHITE

GLASSES

Presented at a conference at the
CUNY Center for Lesbian and Gay
Studies, 9 May 1991.

Today as you passed a dark-skinned
man younger than you
his eyes plucked yours (Indian?
West Indian?),
set you to wondering what he
saw: a close-cropped man
with briefcase, white rolled
shorts, white glasses,

a lope on tiptoes, a
scowl, blue eyes behind the frames,
licking a cone of ice-cream . . .[1]

The first time I met Michael Lynch, I thought his white-framed glasses were the coolest thing I had ever seen. It was at the MLA convention, in New York, in 1986. My first thought was, "Within two months, every gay man in New York is going to be wearing white glasses." My second thought: "Within a year, every fashion-conscious person in the United States is going to be wearing white glasses." My instant resolve: "I want white glasses first."

When I first met Michael Lynch, five years ago in December 1986, it was at an informal coffee shop breakfast meeting to discuss the pos-

---

1. Michael Lynch, "Tobacco," *These Waves of Dying Friends* (New York: Contact II Publications, 1989), p. 65.

sibility of putting out an annual volume of essays in gay and lesbian studies. It wasn't a big breakfast—eight or nine people in a coffee shop—but there was room, at that early though startlingly recent moment in the institutionalization of gay/lesbian literary studies, only five years ago, for almost everyone who had actually published a book in the field, plus a few, like Michael, who hadn't yet. The project was Michael's project because throughout the history of gay studies, of gay activism since Stonewall, Michael has been somebody who has catalyzed, crystallized, fecundated projects, institutions, and communities wherever he has been—wherever geographically, wherever institutionally, wherever in identity and experience.

The day I first met Michael Lynch in New York was the day that, in Toronto, the complicated, arbitrary diagnostic process around AIDS finally caught up with Michael's ex-lover, housemate, best-loved friend, a medical researcher, Bill Lewis, who was to die suddenly the next fall. I heard Michael telling friends around MLA about Bill's diagnosis; I saw the traces of Bill's diagnosis in the amazing dignity and openness with which Michael introduced and chaired a panel I had organized (I, who didn't in 1986 personally know anyone with AIDS) on AIDS and homophobic discourse; the Michael I met and fell in love with was, to some degree I could never estimate, a Michael made different on the same day by the suddenly more graphic proximity of intimate loss—perhaps also by the availability of comfort from friends, and even by the public performance under the MLA's weirdly legitimating and routinizing aegis of an AIDS activist discourse entirely new to that particular theater. Michael's availability to be identified with and loved, in my instant, fetishistic crystallization of him through those white glasses, must have had everything to do with my witness of this moment. At the same time I have always felt, since then, that the important ways in which I *haven't* gotten to know Michael fully are somehow coextensive with my never having known Bill. The same loss, the same history of struggle and subtraction made Michael available to my identification and love, opaque to my knowledge.

And the I who met Michael and fell in love with his white glasses? It was nobody simpler than the handsome and complicated poet and scholar I met in him; it was a queer but long-married young woman whose erotic and intellectual life were fiercely transitive, shaped by a thirst for knowledges and identifications that might cross the barriers of what seemed my identity. It was also someone who had it at heart to make decisive interventions on two scenes of identity that were supposed not to have to do with each other: the scene of feminism, where I "identified" and which I knew well; and the scene of gay men's bonding, community, thought, and politics, a potent and numinous scene which at the experiential level was at that time almost totally unknown to me.

2

Four months ago when I decided to write "White Glasses" for this conference I thought it was going to be an obituary for Michael Lynch. The best thing about writing it is that it isn't—it's an act of homage to a living friend—but someday it will be. I thought I would have to do the speaking of it, and probably the thinking and writing for it, after Michael's death of AIDS-related infection. Which seemed imminent. After months of grogginess, discontinuous attention, extreme weakness, futile attempts to regain weight and alertness, Michael had decided it was time to die: time to end the assaultive doses of antibiotics, to stop stuffing himself with food he didn't want, to take the decision about his fate back into the autonomous hands where it belonged. In making the decision to let himself die, refusing food and all but palliative care, Michael was supported by amazing resources of affection, information, and the most mundane personal care from the communities he had himself created, co-created, and fostered in Toronto. Old and new friends, from *The Body Politic,* from AIDS Action Now!, and from the Canadian Center for Gay and Lesbian Studies, the last of these a new organization Michael had founded in response to Bill Lewis's death, organized a care team for Michael on what I think is an unprecedented model and scale: twenty-four-hour-a-day attendance by a weekly rota of thirty or so friends, organized through Sunday meetings (often with a nurse), instructed and kept scrupulous track of through a massive logbook. . . . Empowerment to decide, permission to die, the knowledge and tending necessary to do so on his own terms—these turned out to have been, not only among the many gifts from the people who love Michael, but a part of Michael's legacy to himself from two decades of activism, writing, and what can only be called the work of community.

3

It took me a year and a half of peering in the window of every optician in New York, northern California, and Massachusetts to find glasses that I thought looked like Michael's. When I got them I felt fashion-perfect— I felt like Michael—even though, in the intervening time, the predicted fashion craze of white glasses had entirely failed to materialize. When I got the glasses I also learned from watching, through them, the faces of other people looking at me, that although to me the glasses meant (mean) nothing but Michael, to others—even to people who know Michael—the glasses don't.

One thing I learned from this is that the white of the glasses means differently for a woman, for a man. The white of the glasses is two things, after all. White is a color—it is a pastel. White the pastel sinks banally and invisibly into the camouflage of femininity, on a woman, a white woman. In a place where it doesn't belong, on Michael, that same pastel remains a flaming signifier.

White is also, however, at the same time no color, the color of color's own subtraction and absence. At once the white-flaring acid of dissolution, the acid's crystalline residue and its voided trail, in many cultures white is the color of mourning. On women of all colors white refers, again banally, to virginity (to virginity as absence or to the absence of virginity) and the flirtations of the veil—to the ways in which our gender tries to construct us heterosexually as absence and as the dissimulating denial of it, and tries also to inscribe in us, as a standard of our own and other people's value, the zero-degree no-color of (not the skin of Europeans themselves but) the abstractive ideology of European domination. A white woman wearing white: the ruly ordinariness of this sight makes invisible the corrosive aggression that white also is: as the blaze of mourning, the opacity of loss, the opacity loss installs within ourselves and our vision, the unreconciled and irreconcilably incendiary energies streaming through that subtractive gap, that ragged scar of meaning, regard, address.

4

When I decided to write "White Glasses" four months ago, I thought my friend Michael Lynch was dying and I thought I was healthy. Unreflecting, I formed my identity as the prospective writer of this piece around the obituary presumption that my own frame for speaking, the margin of my survival and exemption, was the clearest thing in the world. In fact it was totally opaque: Michael didn't die; I wasn't healthy: within the space of a couple of weeks, we were dealing with a breathtaking revival of Michael's energy, alertness, appetite—also with my unexpected diagnosis with a breast cancer already metastasized to several lymph nodes. So I got everything wrong. I thought I knew back then that assigning myself this task in advance, for this conference whose audience I knew would include many of Michael's friends, people who have known him much longer and better than I, as well as people who might never have heard of him and others who know him only as a defining figure in the story of gay studies— I thought it was a good way to deal prospectively and perhaps lucidly with a process of shock and mourning about Michael's loss that had, indeed, already become turbidly disruptive in my life to a degree I found I couldn't share more directly with Michael. Memorials, dedications: places where you say as if to someone else the things you can't say to the people you love. I thought I would have to—I thought I could—address this to you instead of to Michael; and now (yikes) I can do both. The I who does both is also a different one with new fears and temporalities, a newly sharpened appreciation of the love of friends and comrades, new experience of amputation and prosthesis, new knowledge, expectations, angers, luxuries, and dependencies. Now, shock and mourning gaze in both directions through the obituary frame; and much more than shock and mourning; it is exciting that Michael is alive and full of beans today, sick as he is; I think it is exciting to both of us that I am; and in many ways it is full of stimulation and interest, even, to be ill and writing.

5

Now, I know I don't "look much like" Michael Lynch, even in my white glasses. Nobody knows more fully, more fatalistically than a fat woman how unbridgeable the gap is between the self we see and the self as whom we are seen; no one, perhaps, has more practice at straining and straining to span the binocular view between; and no one can appreciate more fervently the act of magical faith by which it may be possible, at last, to assert and believe, against every social possibility, that the self we see can be made visible as if through our own eyes to the people who see us. The stubborn magical defiance I have learned (I *sometimes* feel I have succeeded in learning) in forging a habitable identity as a fat woman is also what has enabled the series of uncanny effects around these white glasses; uncanny effects that have been so formative of my—shall I call it my identification? Dare I, after this half-decade, call it with all a fat *woman*'s defiance, my identity?—as a gay man.

Uncanny effects: effects as of the frame; as of the mask: effects of focal length.

When I am with Michael, often suddenly it will be as if we were fused together at a distance of half an inch from the eye.

Or I will feel as if someone who looked at us might be blinded by the white stigma of our glasses. It sometimes amazes me that anyone can tell us apart.

So often I feel that I see with Michael's eyes—not because we are the same, but because the same prosthetic device attaches to, extends, and corrects the faulty limb of our vision.

It is as if we were both the man in the iron mask; different men in the same iron mask.

When I am in bed with Michael, our white glasses line up neatly on the night table and I always fantasy that I may walk away wearing the wrong ones.

I mention my obsessive imagery to a friend of Michael's, who says, "That's right, I did notice that you and Michael both wear those patio-furniture glasses."

6

"My identity," along with Michael, "as a gay man," I say. Yet our most durable points of mutual reference are lesbian. My favorite picture of Michael was taken in Willa Cather's bed. We are both obsessed with Emily Dickinson. Tokens, readings, pilgrimages, impersonations around Cather, Dickinson, and our other lesbian ego ideals shape and punctuate our history. The first thing Michael did after my diagnosis in February was to bundle into the mail to me a blanket that has often comforted me at his house—a blanket whose meaning to him is its association with the schoolteacher aunt whose bed he used to lie in in childhood, sandwiched in the crack between her and her lifelong companion, wondering whether (after all, he was adopted) it might not be this Boston marriage whose offspring he somehow really, naturally was.

If what is at work here is an identification that falls across gender, it falls no less across sexualities, across "perversions." And across the ontological crack between the living and the dead.

7

Judith Butler in *Gender Trouble:* "Lacan . . . remarks that 'the function of the mask . . . dominates the identifications through which refusals of love are resolved' . . . . In a characteristic gliding over pronominal locations, Lacan fails to make clear who refuses whom. As readers, we are meant, however, to understand that this free-floating "refusal" is linked in a significant way to the mask. If every refusal is, finally, a loyalty to some other bond in the present or the past, refusal is simultaneously preservation as well. The mask thus conceals this loss, but preserves (and negates) this loss through its concealment. The mask has a double function which is the double function of melancholy. The mask is taken on through the process of incorporation which is a way of inscribing and then wearing a melancholic identification in and on the body; in effect, it is the signification of the body in the mold of the Other who has [been] refused.[2] Dominated through appropriation, every refusal fails, and the refuser becomes part of the very identity of the refused, indeed, becomes the psychic refuse of the refused."[3]

Michael Moon on Judith Butler: "I am arguing on behalf of sex-radical groups dealing with AIDS for the desirability, indeed, the necessity . . . of allowing our sex radicalism to pervade our mourning practices to a degree that we have to this point only begun to explore. I believe Butler's analysis of the melancholia of gender has important implications for this possibility. . . . 'Properly' gendered persons, according to Butler's rereading of Freud, are compelled to deny (first to themselves) that they ever felt desire 'inappropriate' to their supposed desire, and are not permitted to grieve over the loss of this 'other desire.' I want to supplement this perception by arguing that 'melancholy,' homo or hetero, is not just about the disavowal and lack of grieving for '*the* other' desire; there are '*many* other' desires— the entire range of 'perversions'—which many people feel compelled to deny and to omit grieving for the loss of. . . . We want to conduct our mourning and grieving in the image of, and as an indispensable part of, this task of collectively and solitarily exploring 'perverse' or stigmatized desire."[4]

8

From a letter to my little brother in the summer of 1987:

"If you leafed through the enclosed snapshots before getting to the prose, you'll have inferred from the unusual prevalence of *white enameled glasses* that we had—last weekend—a lovingly anticipated visit from my Toronto friend Michael, along with his 16-year-old son, Stefan. The visit, very eventless on the surface, probably nice in many ways that will be palpable in

2. Butler writes "has been refused"; I have chosen to bracket the word "been" to record the effects of her notation that "Lacan fails to make clear who refuses whom" (Butler, *Gender Trouble*, pp. 49–50).
3. Ibid.
4. Michael Moon, "Memorial Rags, Memorial Rages," unpublished manuscript. I am very grateful for the opportunity to see this draft and permission to quote from it.

retrospect, seems to be leaving me with a heavy after-turbulence of feeling inadequate and even *bad* as well as impotent, awful, and the other more usual affectionate things.

"I realized as soon as they left Sunday that I had had, in advance, a quite specific fantasy about their visit. This went back to the time three years ago, when our friend Jeff was in the hospital in NYC with the sudden crisis of an autoimmune disease. As we understood it, if he could have made it through that crisis, the disease could probably have been managed and he would have been okay. The whole week or so he was in the hospital before he died, I focused all my energies on the mental image of the time in the coming summer when he, his wife, and his about-to-be-born kid would come up to spend a slow, sunny, maybe impossible weekend lying about on the lawn with us in Amherst—which was going to be the sign that he was fine and everything was fine. Of course, this never happened, but I think I must have kept the fantasy in storage: the feeling that if Michael would only bring Stefan here for a summer weekend and if I could only manage to provide exactly the right background of *dolce far niente,* then nothing could really fail to be (and Michael could not really fail to be) *just fine.*

"In fact, I think the pressure of this not quite recognized fantasy probably made it harder for any of us to relax or connect properly this weekend. But it didn't require that to make a problem, either, probably; the plainest fact seemed to be that Michael, while he tried with his usual sweetness to sparkle and connect, was weak and exhausted, and probably the thing he wanted most—I certainly would have—was to be in his own bed in his own house. The evening they got here he talked about being able to just rest and decompress. But that night he had an upsetting dream. He had to get a lot of dead bodies out of drawers in an office in a high-rise building, put them into coffins, and get them down the elevators and onto the street, where they were going to be carried as part of a funeral-cum-AIDS activist demonstration. But it was Halloween, and the elevators were already full of masked and costumed people; throughout the building was the grim, drunken atmosphere of office parties. Then next day we did the tour of Emily Dickinson's house, which I think he found chillingly alienating and weird (I just registered it as normally alienating and weird, but then I'm used to Amherst). Then, because we wanted to look at a video he'd brought, a CBC special on "AIDS and the Arts" in which he figures as the Poet, we went to a TV rental/repair place in the center of town that was unmistakably something from the twilight zone: a dark burrow, on a second floor that had no intelligible spatial relation to the ordinary blocks of Amherst that supposedly surrounded it, stuffed with hundreds and hundreds of old dead TVs, and inhabited by two guys who, though young, appeared not to have stepped out of the place in fifty years. Michael almost fainted while we were there, and, though he'd slept till noon, slept again for most of the rest of the afternoon. Then he conked out almost immediately after dinner. . . . When they finally left, it felt utterly bitter to let them out of sight or to stop hugging him, maybe because it was only physical touch that ever seemed, all weekend, to burn (Michael feels very hot) through the static of concern, distraction, (his) unpredictable fragility, (my) misplaced bustling, distance (whose?). . . .

"Sentimentality aside, if it ever can be, I've had a hard time processing all this. I have some obvious, even vengefully moralistic things I can tell myself about it (i.e. that things are not fine and not going to be, and in ways that have only minimally to do with any pressure of one's own feelings), but those don't at all answer to the ways in which one does, or may, nonetheless connect—or with the bitterness of not doing so."

9

Last weekend visiting Toronto I had a few minutes to look through the logbook kept by Michael's care team. I leafed back to February, to the time of my diagnosis and mastectomy, and was amazed to find that one caregiver's shift after another had been marked by the restlessness, exhaustion, and pain of Michael's anxiety about what was going on with me in Durham. Of course it didn't surprise me that he was worried about me or compassionated with me, during such a difficult passage. But what I had felt I was experiencing from him at the time—remember that these were the same weeks when Michael was supposed to die, but instead got stronger and stronger—was the tremendous plenitude of the energies he somehow had available to inject into me. The friend who had been, only weeks before, so flickering and disconnected that I worried that it was torturing him to make him talk on the phone, now was at my ear daily with hours of the lore, the solicitude, the ground-level truthtelling and demand for truthtelling that I simply had to have. I also felt miraculously revitalized by the joy of having a real Michael, not the dry-mouthed struggling shade of him, there again to communicate with. I didn't know where these energies came from in Michael—I thought they were produced as if magically by my need of them; to some extent I still think they were—but I see now that they were also carved directly out of Michael's substance, his rest and his peace of mind.

But also a lot of what I needed so unexpectedly to learn from Michael I had already had opportunities to learn. So much about how to be sick—how to occupy most truthfully and powerfully, and at the same time constantly to question and deconstruct, the sick role, the identity of the "person living with life-threatening disease"—had long been embodied in him, and performed by him, in ways which many of us, sick and well, have had reason to appreciate keenly. These are skills that could not have evolved outside the context of liberatory identity politics and AIDS activism, but their flavor is also all Michael's own.[5] I have sometimes condensed them to myself in the unbearably double-edged performative injunction, "Out, out—." As if the horrifying fragility of a life's brief flame could somehow be braced and welded, in the forge of the signifier, as if orthopedically to the galvanizing coming-out imperative of visibility, defiance, solidarity, and self-assertion. From Michael I also seem always to hear the injunction—not the opposite of "Out, out" but somehow a part of it—"Include, include": to entrust as many people as one possibly can with one's actual body and its needs, one's stories about its fate, one's dreams and one's sources of information or hypothesis about disease, cure, consolation, denial, and the state or institutional violence that are also invested in one's illness. It's as though there were transformative political work to be done just by being available to be identified with in the very grain of one's illness (which is to say, the grain of one's own intellectual, emotional, bodily self as refracted through illness and as resistant to it)—being available for identification to friends, but as well to people who don't love one; even to people who may not like one at all nor even wish one well. All of these may nonetheless be brought consciously, even if hatingly, into the world of people living with this disease—just as, whatever one's privilege, a person living with fatal disease in this particular culture is inducted ever more consciously, ever more needily, yet with ever more profound and transformative revulsion into the manglingly differential world of health care under American capitalism. It's been one of the great ideological triumphs of AIDS activism that, for a whole series of overlapping communities, any person living with AIDS is now visible, not only as someone dealing with a particular, difficult cluster of pathogens, but equally as someone who is by that very fact defined as a victim of state violence. What needs to happen now, and I believe can happen, is the even more radical and shaming realization that

5. The best sample of this in writing is Michael Lynch's "Last Onsets: Teaching with AIDS," in *Profession* 1990: 32–36.

under the present regime of systemic exclusion from health care in at least the United States, *every* experience of illness is, among other things, a subjection to state violence, and where possible to be resisted as that.

10

One of the first things I felt when I was facing the diagnosis of breast cancer was, "Shit, now I guess I really must be a woman." A lot of what I was responding to was the way the formal and folk ideologies around breast cancer not only construct it as a secret, but construct it as the secret whose sharing defines women as such. All of this as if the most obvious thing in the world were the defining centrality of her breasts to any woman's sense of her gender identity and integrity! This did not happen to be my situation: as a person who has been nonprocreative by choice, and whose sense of femininity, whatever it may consist in, has never been routed through a pretty appearance in the imagined view of heterosexual men—as a woman moreover whose breast eroticism wasn't strong—I was someone to whom these mammary globes, though pleasing in myself and in others who sported them, were nonetheless relatively peripheral to the complex places where sexuality and gender identity really happen. Something similar has seemed true, moreover, to many other very different women I have talked with, including the other women who sat angrily through a meeting of the hospital-organized breast cancer support group, being told by a social worker that with proper toning exercise, makeup, wigs, and a well-fitting prosthesis, we could feel just as feminine as we ever had and no one (i.e., no man) need ever know that anything had happened. As if our unceasing function is to present, heterosexually, the spectacle of the place where men may disavow their own mortality and need as well as ours. But how different was this, after all, from the message I heard a week before the diagnosis, at an "evening of wit, wisdom, and storytelling for lesbians" by Joanne Loulann, whose hilarious, community-healing, butch/femme-celebrating, powerfully sex-affirmative performance was gored by the ugliness of a single moment when she invidiously compared the paucity of federal research money available to "us" (for research into "our" disease, breast cancer) to the supposed riches being poured into research on AIDS. As though AIDS research were choking on excess of resources!—as though AIDS were *not* a disease of women, of lesbians!—but also as though the identity and solidarity of this lesbian-defined audience depended breathlessly on the intimacy of our association with that-disease-that-is-not-AIDS. As though what we all are, as women, butch, femme, androgynous, alike, is nonetheless not only the thing defined by breasts but also that-thing-that-is-not-man, that is not the male labeled queer, that thing not vulnerable through poverty or racism, through injection, through an insertive or hot and rubbed-raw sexuality to the bad luck of viral transmission.

The invidious comparison has now become a hurtful commonplace, but many of us in the audience were shocked, hearing it then. I feel I must refuse to identify as a woman on this ground. As a woman, I certainly could contract HIV. I happen to have contracted another, also very often fatal disease that makes its own demands of a new politics, a new identity-formation. As a woman, I have been intimately formed by, among other things, the availability for my own identifications of men and of male "perversion," courage, care, loss, struggle, and creativity. I also know that men as well as women have been intimately formed by my and many other women's availability for identification in these ways, and are likely to be even more so in the future. In the day-to-day experience so far of living with and fighting breast cancer, meanwhile, I feel inconceivably far from finding myself at the center of the mysteries of essential femaleness. The people to whose exquisite care I can attribute my present, buoyant spirits and health are the same companions, students, friends, of *every* gender and sexuality, who have always been as vital to my self-formation as I think I have been to theirs. Beyond that, a dizzying array of gender challenges and experiments comes with the initiations of surgery, of chemotherapy, of hormone therapy. Just getting dressed in the morning means deciding how many breasts I will be able to recognize myself if I am wearing (a voice in me keeps whispering, *three*); the apparition of my only slightly fuzzy head, facing me in the mirror after my shower like my own handsome and bald father, demands that I decide if I would feel least alienated or most adventurous or comforted today as Gloria Swanson or Jambi, as a head-covered Hasidic housewife, as an Afro wannabe in a probably unraveling head rag, as a drag queen who never quite figured out how to do wigs, as a large bald baby or Buddha or wise extraterrestrial, or as—my current choice—the befezzed disciple of my new gay fashion gurus, Akbar and Jeff. Indeed, every aspect of a self comes up for grabs under the pressure of modern medicine, with its strange mix of the most delicate and the coarsest of knowledges and imaging capabilities. That pretty, speckled, robin's-egg blue pill with the slightly sinister name "Cytoxan"—it was developed during World War II as a chemical warfare agent; when, as per doctor's instructions, I drop four of this "agent" into my bloodstream every morning, the *mildest* way to describe what is happening is via the postmodernist cliché that I am "putting in question the concept of agency"! I have never felt less stability in my gender, age, and racial identities, nor, anxious and full of the shreds of dread, shame, and mourning as this process is, have I ever felt more of a mind to explore and exploit every possibility.

11

There's so much to be said about the powerfully performative rhetorical force of obituaries and memorials. How can I even think about the task of writing an obituary for someone who's alive? I tell myself sometimes that being sick has made me read obituaries differently, but really I have always been fascinated by them in the same "morbid" way, I have always pro-pelled myself into all the positions around every obituary I saw with the whole force of this particular imagination. My own real dread has never been about dying young but about losing the people who make me want to live. For many other people these things arrange themselves differently; but all are wrung, whirlpooled, turned inside out in the obituary relation. The most compelling thing about obituaries is how openly they rupture the conventional relations of person and of address. From a tombstone, from the tiny print in the *New York Times,* from the panels on panels on panels of the Names Project quilt, whose voice speaks impossibly to whom? From where is this rhetorical power borrowed, and how and to whom is it to be repaid? We miss you. Remember me. She hated to say goodbye. Partici-pating in these speech acts, we hardly know whether to be interpellated as survivors, bereft; as witnesses or even judges; or as the very dead. I look at my snapshots from the 1987 gay and lesbian march on Washington—I see Michael across a distance, white glasses blazing, looking young and forlorn with no mustache; it's only weeks after Bill's sudden death; in the panels of the quilt, I see that anyone, living or dead, may occupy the position of the speaker, the spoken to, the spoken about. When the fabric squares speak, they say,

"Love you! Kelly."

"Frederic Abrams. 'Such Drama.' "

"Roy Cohn. Bully. Coward. Victim."

"Michel Foucault. Where there is power, there is resistance . . . a plu-rality of resistances . . . spread over time and space . . . It is doubtless the strategic codification of these resistances that makes a revolution possible."

"For our little brother, David Lee."

"Sweet dreams."

"Hug Me."

Churned out of this mill of identities crossed by desires crossed by identifications is, it seems—it certainly seemed in October 1987—a fractured and *therefore* militant body of queer rebellion.

But no one can really claim or own the relations of mourning. This winter in North Carolina an ACT UP friend and I went to do some fence-mending with the local Names Project committee—a committee that we sometimes feel is monopolizing, for no purpose more liberatory than memorializing and consolation, the energies and money of a lot of the A-list of North Carolina gays for which we think we could find considerably more telling uses. At what we thought, with relief, was the breakup of a long meeting, the unctuous guy who was chairing the committee announced that to remind ourselves what the committee was all about, and to rededicate ourselves to our (or its) purposes, he and some other people were going to unfold, in the lobby of the building, the latest of the big quilt panels that had passed through their hands—for us to view. I had to do it, but I didn't want to. The quilt wrings me out, as it does any viewer, in a way I don't always want to be available to be wrung out; it was a time when somebody I loved very much was barely hanging on and wasn't sure he wanted to; and just then I was very angry with the project, with its nostalgic ideology and no politics, with its big, ever-growing, and sometimes obstructive niche in the ecology of gay organizing and self-formation. Truculently and furiously I perused, as it unfolded, the random patchwork of other people's mourning, daring it to make me cry (I felt as though I would be lost if I started crying that night)—the now familiar topoi of these terrible losses, the appliquéd rainbows, teddy bears, photographs, baby blocks, 501 jeans; the first-person utterance of survivors, the awful first-person utterance attributed to the dead. I felt, somewhat desperately, as though I knew and could at that moment resist it all. As it turned out, the square I had no way of dealing with was the one appliquéd with SILENCE = DEATH and ACT UP T-shirts: not because of them, but because of the unplaceable, unassuageable voice of its lettering, which said starkly: "HE HATED THE QUILT." I don't know whether my tears and bile were finally those of rage, surrender, envious exultation, or absolute hopelessness. I don't know whose powerful voices I was hearing or which were triumphing: the heroic and now immortalized refusal of the dead man; the—what? I don't know how to begin to characterize the possible tones—of the survivors who honored him and by the same gesture surrendered him up to what he hated; or the ravenously denuding, homogenizing, relentlessly anthropomorphizing and yet relentlessly disorienting abyssal voice of the obituary imperative, the implacably inclusive format of memorial relation and address.

12

I think Michael is very, very tired of being sick, and I think I can feel that with him—though I also feel that every day that Michael is there and recognizably himself, gossipy, courageous, universally inquisitive, perhaps crabby, communicative, and craving physical touch, is a day that I have an important reason to be happy. My own illness hasn't really even begun to come home to roost—it probably won't for some years, *maybe* never—but I also see, or imagine, some of the people who love me beginning to deal with the possibility that someday the same calculus may operate around my own fatigue, discouragement, pain, flares of zest and creativity, the recognizable, recognizing, and hurting shards of relation and identity. I still want to know more and more about how Michael and other people deal with this long moment, and about how I will. As whom, as what I may deal with it; out of what spaces I may speak of it, or be spoken for in these identities and struggles—I know these are not simply for me or even for my immediate communities to decide; yet I relish knowing that enough of us will be here to demonstrate that the answer can hardly be what anyone will have expected.

13

A week ago at a country inn on Lake Huron, Michael and another friend and I were talking about White Glasses ("White Glasses" the talk, not white glasses the glasses). Michael said—perhaps apprehensively—"I'm certainly glad I'm not going to be there."

I said—*distinctly* apprehensively—"I'm certainly glad you're not going to be there too. But I still want you to get a kick out of this."

"Oh I *am,* I am," Michael said. "Are you going to record it for me?" So I am recording it. Hi Michael! I know I probably got almost everything wrong but I hope you didn't just hate this. See you in a couple of weeks.

Michael Lynch died of AIDS on 9 July 1991.

BIBLIOGRAPHY

Adler, Jerry, et al. "Taking Offense: Is This the New Enlightenment on Campus or the New McCarthyism?" *Newsweek*, 24 December 1990, pp. 48–55.

Anderson, Benedict. *Imagined Communities: Reflections on the Origin and Spread of Nationalism*. London: Verso Press, 1983.

Anzaldúa, Gloria. *Borderlands/La Frontera*. San Francisco: Spinsters/Aunt Lute press, 1989.

Assézat, J. *Oeuvres complètes de Diderot*. 5 vols. Paris: Garnier Frères, 1875.

Austen, Jane. *Sense and Sensibililty*. Harmondsworth, Middlesex: Penguin Books, 1967.

Barker-Benfield, G. J. *The Horrors of the Half-Known Life: Male Attitudes Toward Women and Sexuality in Nineteenth-Century America*. New York: Harper and Row, 1976.

Bartlett, Neil. *Who Was That Man?: A Present for Mr. Oscar Wilde*. London: Serpent's Tail, 1988.

Bell, A. P., M. S. Weinberg, and S. K. Hammersmith. *Sexual Preference: Its Development in Men and Women*. Bloomington: Indiana University Press, 1981.

Bennett, Paula. *Emily Dickinson: Woman Poet*. Iowa City: University of Iowa Press, 1990.

———. *My Life A Loaded Gun: Female Creativity and Feminist Poetics*. Boston: Beacon Press, 1986.

Bernstein, Richard. "Academia's Liberals Defend Their Carnival of Canons Against Bloom's 'Killer B's.' " *New York Times*, 25 September 1988, p. 26E.

———. "The Rising Hegemony of the Politically Correct: America's Fashionable Orthodoxy." *New York Times*, 28 October 1990, sec. 4 (Week in Review), pp. 1, 4.

Berridge, Virginia, and Griffith Edwards. *Opium and the People: Opiate Use in Nineteenth-Century England*. 2d ed. New Haven, Conn.: Yale University Press, 1987.

Bersani, Leo. *The Culture of Redemption*. Cambridge, Mass.: Harvard University Press, 1990.

———. "The Culture of Redemption: Marcel Proust and Melanie Klein." *Critical Inquiry* 12 (Winter 1986): 399–421.

———. "Is the Rectum a Grave?" In Douglas Crimp, ed., *AIDS: Cultural Analysis, Cultural Activism*, pp. 197–222. Cambridge, Mass.: MIT Press, 1988.

Boston Women's Health Book Collective. *The New Our Bodies, Ourselves: Updated and Expanded for the Nineties*. New York: Simon and Schuster, 1992.

Brady, Judy, ed. *1 in 3: Women with Cancer Confront an Epidemic*. Pittsburgh and San Francisco: Cleis Press, 1991.

Bridenthal, Renate, Atina Grossmann, and Marion Kaplan, eds. *When Biology Became Destiny: Women in Weimar and Nazi Germany*. New York: Monthly Review Press, 1984.

Brody, Jane E. "Personal Health." *New York Times*, 4 November 1987.

Brontë, Charlotte. *Jane Eyre.* 2 vols. Edinburgh: John Gray, 1905.

―――. *Villette.* Edinburgh: John Gray, 1905.

Brownmiller, Susan. *Against Our Will: Men, Women, and Rape.* New York: Simon and Schuster, 1975.

Bull, Chris. "No More Dirty Waters." *Advocate* 549 (24 April 1990): 30–35.

Butler, Judith. *Gender Trouble: Feminism and the Subversion of Identity.* New York: Routledge, 1990.

Butler, Sandra, and Barbara Rosenblum. *Cancer in Two Voices.* San Francisco: Spinsters, 1991.

Califia, Pat. *Sapphistry: The Book of Lesbian Sexuality.* Tallahassee, Fla.: Naiad Press, 1983.

Caplan, Jane. "Introduction to Female Sexuality in Fascist Ideology." *Feminist Review* 1 (1979).

Caplan, Jay. *Framed Narratives: Diderot's Genealogy of the Beholder.* (Afterword by Jochen Schulte-Sasse.) Theory and History of Literature, vol. 19. Minneapolis: University of Minnesota Press, 1985.

Case, Sue-Ellen. "Tracking the Vampire." *differences: A Journal of Feminist Cultural Studies* 3, no. 2 (1991): 1–20.

Cather, Willa. *Five Stories.* New York: Vintage Books, 1956.

―――. *The Kingdom of Art: Willa Cather's First Principles and Critical Statements, 1893–1896.* Edited by Bernice Slote. Lincoln: University of Nebraska Press, 1966.

―――. *The Professor's House.* New York: Vintage Books, 1973.

―――. *The Troll Garden.* New York: New American Library, 1971.

Cavell, Stanley. *Must We Mean What We Say?* New York: Scribners, 1969.

Cohen, Ed. *Talk on the Wilde Side.* New York: Routledge, forthcoming.

Craft, Christopher. "Alias Bunbury: Desire and Termination in *The Importance of Being Earnest.*" *Representations* 31 (Summer 1990): 19–46.

Daniels, Lee A. "Liberal Arts Scholars Seek to Broaden Their Field." *New York Times,* 21 September 1988, p. 23.

de la Carrera, Rosalina. "Epistolary Triangles: The *Préface-Annexe* of *La Religieuse* Reexamined." *The Eighteenth Century: Theory and Interpretation* 29 (Winter 1988): 263–80.

de Lauretis, Teresa. "Sexual Indifference and Lesbian Representation." *Theatre Journal* 40 (May 1988): 155–77.

D'Emilio, John, and Estelle B. Freedman. *Intimate Matters: A History of Sexuality in America.* New York: Harper and Row, 1988.

Diagnostic and Statistical Manual of Mental Disorders. 3d ed. Washington, D.C.: American Psychiatric Association, 1980.

Dickens, Charles. *The Personal History of David Copperfield.* Edited by Trevor Blount. Harmondsworth, Middlesex: Penguin Books, 1966.

Dickinson, Emily. *The Complete Poems of Emily Dickinson.* Edited by Thomas H. Johnson. Boston: Little, Brown, 1960.

Diderot, Denis. *The Nun.* Translated by Leonard Hancock. Harmondsworth, Middlesex: Penguin Books, 1974.

Dollimore, Jonathan. *Sexual Dissidence: Augustine to Wilde, Freud to Foucault.* New York: Oxford University Press, 1991.

Douglas, Mary. *Purity and Danger.* New York: Praeger Books, 1967.

Drabble, Margaret, ed. *The Oxford Companion to English Literature.* 5th ed. New York: Oxford University Press, 1985.

Faderman, Lillian. *Surpassing the Love of Men: Romantic Friendship and Love between Women from the Renaissance to the Present*. New York: William Morrow, 1981.

Felman, Shoshana. *The Literary Speech Act: Don Juan with J. L. Austin, or Seduction in Two Languages*. Translated by Catherine Porter. Ithaca, N.Y.: Cornell University Press, 1983.

Ferenczi, Sandor. "Confusion of Tongues between Adults and the Child: The Language of Tenderness and of Passion." In Michael Balint, ed., *Final Contributions to the Problems and Methods of Psycho-analysis*, pp. 156–67. New York: Basic Books, 1955.

Fineman, Joel. "The Significance of Literature: *The Importance of Being Earnest*." *October* 15: 79–90.

Forster, E. M. *The Longest Journey*. Edited by Elizabeth Heine. London: Penguin Books, 1984.

——— . *Maurice*. New York: W. W. Norton, 1981.

——— . *A Passage to India*. San Diego: Harcourt Brace Jovanovich, 1984.

Foucault, Michel. *The History of Sexuality: An Introduction*. Translated by Robert Hurley. New York: Pantheon Books, 1978.

Fried, Michael. *Absorption and Theatricality: Painting and Beholder in the Age of Diderot*. Berkeley: University of California Press, 1980.

——— . "Realism, Writing, and Disfiguration in Thomas Eakins's *Gross Clinic*." *Representations* 9 (Winter 1985): 33–104.

Friedman, Richard C. *Male Homosexuality: A Contemporary Psychoanalytic Perspective*. New Haven, Conn.: Yale University Press, 1988.

Freud, Sigmund. "A Child Is Being Beaten." *Collected Papers of Sigmund Freud*. Authorized translation under the supervision of Joan Rivière. 5 vols. New York: Basic Books, 1959.

Fung, Richard. *Looking for My Penis*. Video. 1989.

Gallagher, Catherine. "The Body Versus the Social Body in the Works of Thomas Malthus and Henry Mayhew." In Catherine Gallagher and Thomas Laqueur, eds., *The Making of the Modern Body: Sexuality and Society in the Nineteenth Century*, pp. 83–106. Berkeley: University of California Press, 1987.

Garber, Marjorie. *Vested Interests: Cross-Dressing and Cultural Anxiety*. New York: Routledge, 1992.

Gibson, Paul. "Gay Male and Lesbian Youth Suicide." In U.S. Department of Health and Human Services, *Report of the Secretary's Task Force on Youth Suicide*, vol. 3, pp. 110–42. Washington, D.C., 1989.

Goldberg, Rita. *Sex and Enlightenment: Women in Richardson and Diderot*. Cambridge: Cambridge University Press, 1984.

Grahn, Judy. *Another Mother Tongue: Gay Words, Gay Worlds*. Boston: Beacon Press, 1984.

Green, Richard. *The "Sissy Boy Syndrome" and the Development of Homosexuality*. New Haven, Conn.: Yale University Press, 1987.

Halperin, David M. *One Hundred Years of Homosexuality*. New York: Routledge, 1989.

Harding, Rick. "Sex Education in Washington, D.C.: Dannemeyer Talks Dirty on the Floor of Congress." *Advocate* 534 (26 September 1989): 10.

Hare, E. H. "Masturbatory Insanity: The History of an Idea." *Journal of the Mental Sciences* 108 (1962): 1–25.

Hertz, Neil. *The End of the Line*. New York: Columbia University Press, 1985.

Hollander, Anne. *Seeing Through Clothes*. New York: Viking Press, 1978.

Holleran, Andrew. *Dancer from the Dance*. New York: William Morrow, 1978; Bantam, 1979.

James, Henry. *The Golden Bowl.* Harmondsworth, Middlesex: Penguin Books, 1973.

———. *Notebooks of Henry James.* Edited by F. O. Matthiessen and Kenneth B. Murdock. New York: Oxford University Press, 1947.

———. *The Wings of the Dove.* Edited by John Bayley. London: Penguin Books, 1986.

Julien, Isaac. *Looking for Langston.* Film. 1989.

Kellogg, T. A., et al. "Prevalence of HIV-1 Among Homosexual and Bisexual Men in the San Francisco Bay Area: Evidence of Infection Among Young Gay Men." In *Seventh International AIDS Conference Abstract Book,* vol. 2 (Geneva, 1991) (W.C. 3010), p. 298.

Kimball, Roger. *Tenured Radicals: How Politics has Corrupted Our Higher Education.* New York: Harper and Row, 1990.

Kipling, Rudyard. *The Writings in Prose and Verse of Rudyard Kipling.* 35 vols. New York, 1907.

Koestenbaum, Wayne. *Double Talk: The Erotics of Male Literary Collaboration.* New York: Routledge, 1989.

Kramarae, Cheris, and Paula A. Treichler. *A Feminist Dictionary.* Boston: Pandora Press, 1985.

Laplanche, Jean. *New Foundations for Psychoanalysis.* Translated by David Macey. Oxford: Basil Blackwell, 1989.

Lewes, Kenneth. *The Psychoanalytic Theory of Male Homosexuality.* New York: Simon and Schuster, 1988; Penguin/NAL/Meridian, 1989.

Lister, Anne. *I Know My Own Heart: The Diaries of Anne Lister (1791–1840).* Edited by Helena Whitbread. London: Virago, 1988.

Lorde, Audre. *A Burst of Light.* Ithaca, N.Y.: Firebrand Books, 1988.

———. *The Cancer Journals.* 2d ed. San Francisco: Spinsters Ink, 1988.

Lynch, Michael. "Last Onsets: Teaching with AIDS." *Profession* (1990): 32–36.

———. *These Waves of Dying Friends.* New York: Contact II, 1989.

McConnell-Ginet, Sally. "The Sexual (Re)Production of Meaning: A Discourse-Based Theory." Manuscript.

MacDonald, Robert H. "The Frightful Consequences of Onanism: Notes on the History of a Delusion." *Journal of the History of Ideas* 28 (1967): 423–31.

Mass, Lawrence. *Dialogues of the Sexual Revolution,* vol. 1: *Homosexuality and Sexuality.* New York: Harrington Park Press, 1990.

Masson, Jeffrey Mousaieff. *A Dark Science: Women, Sexuality, and Psychiatry in the Nineteenth Century.* New York: Farrar, Straus and Giroux, 1986.

Miller, Alice. *For Your Own Good: Hidden Cruelty in Child-Rearing and the Roots of Violence.* Translated by Hildegarde Hannum and Hunter Hannum. New York: Farrar, Straus and Giroux, 1983.

———. *Thou Shalt Not Be Aware: Psychoanalysis and Society's Betrayal of the Child.* Translated by Hildegarde Hannum and Hunter Hannum. New York: Farrar, Straus and Giroux, 1984.

Miller, D. A. "Anal Rope." In Diana Fuss, ed., *Inside/Out: Lesbian Theories, Gay Theories,* pp. 119–41. New York: Routledge, 1991.

———. *The Novel and the Police.* Berkeley: University of California Press, 1988.

Mitchell, Juliet. *Psychoanalysis and Feminism.* New York: Pantheon Books, 1974; Vintage Books, 1975.

Money, John. *The Destroying Angel: Sex, Fitness and Food in the Legacy of Degeneracy*

Theory, Graham Crackers, Kellogg's Corn Flakes and the American Health History. Buffalo, N.Y.: Prometheus, 1985.

Moon, Michael. "Memorial Rags, Memorial Rages." Manuscript.

———. "Sexuality and Visual Terrorism in *The Wings of the Dove.*" *Criticism* 28 (Fall 1986): 427–43.

———. "A Small Boy and Others: Sexual Disorientation in Henry James, Kenneth Anger, and David Lynch." In Hortense J. Spillers, ed., *Comparative American Identities: Race, Sex, and Nationality in the Modern Text,* pp. 141–56. New York: Routledge, 1991.

Moore, Rayburn S., ed. *Selected Letters of Henry James to Edmund Gosse, 1882–1915: A Literary Friendship.* Baton Rouge: Louisiana State University Press, 1988.

Mosse, George L. *Nationalism and Sexuality: Respectability and Abnormal Sexuality in Modern Europe.* New York: Fertig, 1985.

Mullan, John. *Sentiment and Sociability: The Language of Feeling in the Eighteenth Century.* New York: Oxford University Press, 1988.

Mylne, Vivienne. "What Suzanne Knew: Lesbianism and *La Religieuse.*" *Studies on Voltaire and the Eighteenth Century* 208 (1982): 167–73.

Nardin, Jane. "Children and Their Families." In Janet Todd, ed., *Jane Austen: New Perspectives,* Women and Literature, New Series, vol. 3, pp. 73–87. New York: Holmes and Meier, 1983.

Neuman, Robert P. "Masturbation, Madness, and the Modern Concept of Childhood and Adolescence." *Journal of Social History* 8 (1975): 1–22.

Owens, Craig. *Beyond Recognition: Representation, Power, and Culture.* Berkeley: University of California Press, 1992.

Paglia, Camille. "Oscar Wilde and the English Epicene." *Raritan* 4 (Winter 1985): 85–109.

———. *Sexual Personae: Art and Decadence from Nefertiti to Dickinson.* New Haven, Conn.: Yale University Press, 1990.

Patton, Cindy. *Inventing AIDS.* New York: Routledge, 1990.

Pear, Robert. "Rights Laws Offer Only Limited Help on AIDS, U.S. Rules." *New York Times,* 23 June 1986, pp. 1, 13.

Plant, Richard. *The Pink Triangle: The Nazi War Against Homosexuals.* New York: Henry Holt, 1986.

Pope, Alexander. *The Poems of Alexander Pope.* Edited by John Butt. New Haven, Conn.: Yale University Press, 1963.

Proust, Marcel. "La Race des tantes." In *"Contre Sainte-Beuve" suivi de "Nouveaux mélanges,"* 14th ed. Paris: Gallimard, 1954.

———. *Remembrance of Things Past.* Translated by C. K. Scott Moncrieff and Terence Kilmartin. 3 vols. New York: Vintage Books, 1982.

Remafedi, G. "Male Homosexuality: The Adolescent's Perspective." Unpublished manuscript, Adolescent Health Program, University of Minnesota. 1985.

Rex, Walter E. "Secrets from Suzanne: The Tangled Motives of *La Religieuse.*" *The Eighteenth Century: Theory and Interpretation* 24 (Fall 1983): 185–98.

Rosario, Vernon A., II. "The 19th-Century Medical Politics of Self-Defilement and Seminal Economy." Paper presented at the Center for Literary and Cultural Studies, Harvard University, "Nationalisms and Sexualities" conference, June 1989.

Rosenblatt, Roger. "The Universities: A Bitter Attack . . ." *New York Times Book Review,* 22 April 1990, p. 3.

Rubenstein, Hal. "Simply Divine." *Interview* 18 (February 1988): 51.

Rubin, Gayle. "The Traffic in Women: Notes on the 'Political Economy' of Sex." In Rayna R. Reiter, ed., *Toward an Anthropology of Women*, pp. 157–210. New York: Monthly Review Press, 1975.

Schor, Naomi. *Reading in Detail: Aesthetics and the Feminine.* New York: Methuen, 1987.

Sedgwick, Eve Kosofsky. *Between Men: English Literature and Male Homosocial Desire.* New York: Columbia University Press, 1985.

———. *Epistemology of the Closet.* Berkeley: University of California Press, 1990.

———. "Trace at 46." *Diacritics* 10 (March 1980): 11–12.

Sherman, Nick. "Fighting Words" (interview with Michael Lynch). *Xtra!* 138 (8 December 1989): 7.

Shorter, Clement. *The Brontës: Life and Letters.* 2 vols. New York, 1908.

Showalter, Elaine. "Critical Cross-Dressing: Male Feminists and the Woman of the Year." In Alice Jardine and Paul Smith, eds. *Men in Feminism*, pp. 116–32. New York: Methuen, 1987.

———. *The Female Malady: Women, Madness, and English Culture, 1830–1980.* New York: Pantheon Books, 1985.

Silverman, Kaja. *Male Subjectivity at the Margins.* New York: Routledge, 1992.

———. "Too Early/Too Late: Subjectivity and the Primal Scene in Henry James." *Novel* 21 (Winter/Spring 1988): 147–73.

Smith-Rosenberg, Carroll. *Disorderly Conduct: Visions of Gender in Victorian America.* New York: Oxford University Press, 1985.

Solomon, Alisa. "The Politics of Breast Cancer." *Village Voice*, 14 May 1991, pp. 22–27.

Stein, Judith, and RaeRae Sears. "Fat, Lesbian, and Proud." Coralville, Iowa: Fat Liberator Publications, n.d.

Stengers, Jean, and Anne van Neck. *Histoire d'une grande peur: la masturbation.* Brussels: Éditions de l'Université de Bruxelles, 1984.

Stocker, Midge, ed. *Cancer as a Women's Issue: Scratching the Surface.* Chicago: Third Side Press, 1991.

Stoller, Robert. *Sexual Excitement.* New York: Pantheon Books, 1979.

Tanner, Tony. *Jane Austen.* Cambridge, Mass.: Harvard University Press, 1986.

Untermeyer, Louis, ed. *The Golden Treasury of Poetry.* New York, 1959.

Waters, John. *Crackpot: The Obsessions of John Waters.* New York: Vintage Books, 1987.

———. *Shock Value: A Tasteful Book About Bad Taste.* New York: Delta Books, 1981.

———. *Trash Trio: Three Screenplays:* Pink Flamingos, Desperate Living, *and* Flamingos Forever. New York: Vintage Books, 1988.

Weeks, Jeffrey. *Coming Out: Homosexual Politics in Britain, from the Nineteenth Century to the Present.* London: Quartet Books, 1977.

Wilde, Oscar. The Importance of Being Earnest *and Other Plays.* Introduction by Sylvan Barnet. New York: New American Library, 1985.

———. *The Importance of Being Earnest. A Trivial Comedy for Serious People. In Four Acts as Originally Written.* 2 vols. New York: New York Public Library, 1956.

———. *The Picture of Dorian Gray.* Harmondsworth, Middlesex: Penguin Books, 1981.

Wilner, Joshua. "Autobiography and Addiction: The Case of De Quincey." *Genre* 14 (Winter 1981): 493–503.

———. "The Stewed Muse of Prose." *MLN* (December 1989): 1085–98.

Wojnarowicz, David. *Close to the Knives: A Memoir of Disintegration.* New York: Vintage Books, 1991.

Zambaco, Démétrius. "Onanism and Nervous Disorders in Two Little Girls." Translated by Catherine Duncan. *Semiotext(e)* 4, no. 1 ("Polysexuality") (1981): 22–36.

INDEX

ACT UP (AIDS Coalition to Unleash Power), xi, 14

Addiction: history of the term, 109–13, 123–25, 130–35; HIV/AIDS and, 136; homosexual identity and, 135–36, idea of the "will" and, 123–24, 132–42; masturbation and, 122

AIDS. *See* HIV/AIDS

AIDS memorials, 104–6, 252–66

Althusser, Louis, 97

Anal eroticism: AIDS discourse and, 210n; female, 177–78, 203, 211; gender identity and, 203; James and, 96–103; male, 203–4; male-male rape and, 247–48; poetry and, 177–214; representations of modern homosexual identity and, 203n; Wilde and, 67–68. *See also* Anus; Sodomy

Anderson, Benedict, 144–49

Anger, Kenneth, 237

Anglund, Joan Walsh, 181

Anorexia, 131

Anti-intellectualism, 17–20

Anti-pornography movement, 188n

Anti-Semitism, 49n

Anus: AIDS and representations of, 246–47; as alternative to phallic economy, 98–103, 245–49; representations of James, 96–103; representations of in Waters's films, 245–47. *See also* Anal eroticism

Anzaldúa, Gloria, 9

Assézat, J., 26–29

Aunt. *See* Avunculate

Austen, Jane, xiii, 16; *Sense and Sensibility*, 109–29

Austin, J. L., 9

Autoeroticism: in Austen, 109–29; in Dickinson, 115; in James, 93–95, 99–103; sexual identity and, 8, 111–13, 116–18. *See also* Binary oppositions; Masturbation

Avunculate: as queer affiliation, 62–63; in Wilde, 52–72

Balint, Michael, 64n

Barker, Francis, 247

Barker-Benfield, G. J., 116n

Barrie, James M., 90

Barry, Linda, 84

Bartlett, Neil, 59n

Bell, A. P., 158n

Bennett, Paula, 115

Bernstein, Richard, 109n, 146n

Berridge, Virginia, 123, 130

Berry, Chuck, 20

Bersani, Leo, 70, 174, 246

Binary oppositions: active/passive, 101; autoerotic/alloerotic, 101, 109–13, 116, 136; constructivism/essentialism, 158–64, 226–27; crossings of homo/heterosexual, 57–58, 171–74; crossings of male/female, 171–74; fisting as switchpoint between, 101; heterosexual/male violence, 137; homosexual/heterosexual, 101, 115–17; homosocial/homosexual, 49–51; immanent/extrinsic, 137; innocence/ignorance, 37–43; knowledge/ignorance, 23–51;